POLLUTION PREVENTION OPPORTUNITY ASSESSMENTS

POLLUTION PREVENTION OPPORTUNITY ASSESSMENTS

A Practical Guide

MARCUS J. HEALEY, PH.D.
DANIEL WATTS, PH.D.

Additional Authors
LAURA L. BATTISTA
THOMAS GROSVENOR
DOUGLAS A. KRETKOWSKI
JEFFREY LEWIS
JEFFREY D. PICO

JOHN WILEY & SONS, INC.
New York • Chichester • Weinheim • Brisbane • Singapore • Toronto

Published simultaneously in Canada.

Library of Congress Cataloging-in-Publication Data:

Pollution prevention opportunity assessments : a practical guide /
Marcus J. Healey . . . [et al.].
p. cm.
Includes bibliographical references and index.
ISBN 0-471-29226-5
1. Industrial management—Environmental aspects. 2. Pollution—Economic aspects. I. Healey, Marcus J.
HD30.255.p66 1997
658.4'08—dc21 97-37397
CIP

10 9 8 7 6 5 4 3 2 1

Contents

Preface

The environmental field is undergoing rapid change. At a time when we have begun to appreciate the meaning of terms such as pollution control and waste minimization, new terms such as pollution prevention, total cost accounting, life-cycle assessment, sustainable development, and industrial ecology are added to the list. The result is a new perspective, a fount of new knowledge to foster and encourage new insights. *Pollution Prevention Opportunity Assessments: A Practical Guide* focuses on topics of interest for today's environmental professional. This text is designed to provide the reader with a fundamental understanding of the subject and, it is hoped, to foster a desire for additional training and information.

Environmental professionals of all stripes, as well as managers, engineers, technicians, and trainers should obtain much benefit from the information contained in this text. Those readers that require or perhaps only desire a practical resource to help guide them in pollution prevention planning may find this a valuable source of information. It is also hoped that those professionals currently involved in the environmental arena might find this material somewhat enjoyable and, at the very least, refreshing. Other professionals who are not directly involved with environmental issues, but who must work with those who are, will also find this text useful.

Those practitioners who desire more instruction in this most complex subject should visit our website at www.njit.edu/njtap.

Marcus J. Healey
Newark, New Jersey

Acknowledgments

I wish to acknowledge the individuals and organizations who have assisted in the development of this publication. First, I would like to thank my family for their encouragement and patience. Second, I would like to thank Laura Lynne Battista for her unswerving devotion, beyond the call of duty, to making this publication a reality. Third, thanks to Dr. Bruce Herrick for his invaluable advice, suggestions, and input. Finally, thanks to the faculty and staff at New Jersey Institute of Technology who provided unflinching support and incisive advice.

I wish to thank the other authors who persevered in bringing this project to completion. I appreciate their willingness to take time from their busy schedules to help with this effort. It is difficult to put together a publication of this scale. I hope that seeing this book materialize encourages these authors, all experts in their respective fields, to continue writing and publishing. We have only begun to explore the potential of pollution prevention and the new environmental wave that further encompasses life cycle analysis, industrial ecology, and sustainable development. I look forward to our continued journey together. Thanks to all of you. And a special thank you to the superb staff at Van Nostrand Reinhold for their support and the opportunity they made possible.

POLLUTION PREVENTION OPPORTUNITY ASSESSMENTS

1 Introduction

Running a successful business is tough, and evaluating all of the new management trends may seem overwhelming. Voices from many sides talk about the merits of new management approaches, "right-sizing," reengineering, total quality management, and rigorous enforcement of regulations, among other issues. Yet the ultimate avenue for success in business continues to be providing a product or service to customers at an affordable price that still gives appropriate levels of return to the business.

Simultaneously with these other developments, discussions about environmental management, new approaches to environmental regulation, and approaches to continued improvement of the environment have been taking place. The reliance on "end-of-the-pipe" treatment as a regulatory approach has produced substantial improvement in the quality of the air and water and in waste management practices. Yet the cost has been high, and the cost predictions to achieve the environmental quality levels deemed necessary in the future are even higher.

The high cost, coupled with limited results, has lead business people and environmental policy makers to consider alternative routes to address the environmental expectations of the nation's population and its laws. They have considered the reality of the costs required to meet them as well. Over the last few years several new ideas have been advanced and implemented to differing degrees. The various approaches have been described in the federal waste management hierarchy, which essentially states that:

- The best waste management approach is to prevent waste at its source.
- If that cannot be done, then recover, reuse, or recycle it.
- If that cannot be done, then treat the waste in an environmentally appropriate way.
- Any residues or residuals from the treatment process should be disposed of safely.

In most cases, research and implementation started at the bottom of the hierarchy. Much more effort was spent on disposal and treatment needs than on source reduction and recycling. In fact, more attention was given generally to recycling than to source reduction. This level of attention has provided substantial environmental benefit and improvement. Yet as progress was made it seemed clear that more attention to reduction of waste at the source should result in both environmental and economic benefits. Costs should be saved from materials formerly wasted that now could be put to beneficial use, and costs associated with recycling, treatment, and disposal would be avoided. In addition, regulatory costs, such as permits, sampling, and lab fees, could be bypassed.

Also clearly realized was the fact that regulation could force changes in waste disposal and in treatment technology and application, and that incentives, such as deposit laws and tax credits, could be developed to encourage recycling and reuse. However, forcing source reduction was much more difficult because it nearly always required modifications in industrial processes or products.

Some states have implemented pollution prevention laws that require pollution prevention planning. This approach is based on the assumption that when a company develops an effective pollution prevention plan, the environmental and economic benefits will be so apparent that implementation of pollution prevention options will start quickly.

Other states have developed toxic use reduction laws that specify elimination or severe curtailment of the use of certain specific chemicals. There is, however, no particular replacement chemical specified. It is up to the individual companies to identify replacements and implement their use. Source reduction, therefore, involves examination of existing manufacturing processes and product designs in a new light—looking specifically for parts of the operation that produce waste—and identifying possible changes that can reduce or eliminate each individual waste stream.

Pollution prevention can result in economic payoff with regard to avoiding costs of raw materials and of treatments. Other possible benefits include:

- More-efficient operations
- Increased workplace safety

- Avoidance of future liability
- Easier compliance with environmental regulations

The purpose of this book is to provide practical steps and hints for planning and implementing a pollution prevention program at a manufacturing facility. Several types of information are presented illustrating different approaches to the development of such a program. One section provides details of a procedure for gathering information, setting pollution prevention priorities, developing options, and instituting a program. At the first reading, this may appear to be a complicated and difficult procedure. However, the process (or others closely related to it) has been carried out at hundreds of facilities with positive results. As indicated later in the book, it may be effective to start with a smaller demonstration pollution prevention project and use that as a basis for building a broader program.

Other sections of the book provide accounts of specific pollution prevention opportunities that have been implemented at facilities in several different industrial categories. Some of these changes may be readily adapted for use at other facilities. In other cases they may spur thinking that helps identify specific opportunities for other industries or special manufacturing situations.

One section of the book specifically identifies several types of operations such as cleaning and degreasing that are practiced in several industrial sectors. Changes related to pollution prevention for these process functions can have wide positive impact. This illustrates the fact that pollution prevention teams can benefit from examination of other programs in seemingly unrelated industries because common operations may exist.

The book concludes with discussion about available sources of information and ideas about development and implementation of pollution prevention programs. Electronic information sources through the Internet and the World Wide Web are becoming increasingly common and more helpful. Federal agencies, including the EPA and the Department of Energy, offer pollution prevention information. Several states have established pollution prevention technical assistance programs that will usually make site visits to help initiate or improve a pollution prevention program. Similarly, many trade and professional organizations have assistance programs available.

Pollution prevention activities at a facility should be seen as a permanent program, moving toward continued environmental improvement. A one-time examination and set of changes will have some benefit, but regular review and change over a period of time can be expected to lead to increased savings and positive changes in efficiency and product quality. Programs work most effectively when there is real commitment and participation by all employees. It should be recognized, however, that this wide level of buy-in may not be achieved immediately.

Just as pollution prevention at a facility can be seen as an evolving activity, ideas about the goals for environmental management for the nation are evolving. Some of the newer ideas address issues such as "sustainable development" and "industrial ecology." These concepts consider environmental management more broadly than operations at a single manufacturing facility. Industrial ecology considers flows of energy and materials from one manufacturing facility to another, ultimately to the consumer, and then through recycling, reuse, or disposal. The goal of the approach is to borrow from nature concepts such as symbiosis, multistep use and degradation, and nutrient cycling to try to mimic the "tight" material cycles of mature ecosystems. As ecosystems mature they begin to use resources more efficiently. Industrial ecology is a component of sustainable development, which has several possible definitions. The easiest definition says that all aspects of production, use, reuse, and disposal of goods should be carried out in such a way that future generations are not impacted negatively. Among other things, this implies that there will be no overuse of raw materials that cannot be regenerated and that no portion of the environment—air, water, or land—will be degraded.

It is not clear how these newer ideas will enter into environmental regulation. However, it is clear that a facility that starts and continually improves a pollution prevention program will contribute to achieving sustainable development. The process described in this book is not a recipe, nor is it a road map. In some ways it is an outline, or perhaps a set of footprints painted on the floor that set off in an appropriate direction with the expectation that the walker will gain confidence and experience and continue the journey.

Pollution prevention programs have been implemented before. There is an abundance of information about approaches that have worked and others that have been less successful. Opportunities abound for progress and new accomplishments. It is the hope of the authors that this book will assist new pollution prevention practitioners in recognizing and realizing these opportunities.

2

Standard Operating Procedures in the Workplace

How to Establish a Pollution Prevention Program in the Workplace

Industrial waste generation in the United States has reached daily averages in the billions of pounds. This waste discharged to the air, water, or land represents a significant loss of raw materials and a potential threat to human health and the environment. To be effective business managers with an eye on the bottom line as well as responsible guardians of environmental quality, all waste generators should review production processes and operations, and consider both the economic and the environmental benefits of implementing a pollution prevention program.

Adopting a pollution prevention program as a way of doing business can provide significant benefits to a company. By decreasing the amount of waste generated or released, a company can reduce waste disposal costs, improve worker safety, and reduce long-term liability. In addition, pollution prevention methods often increase the efficiency of the production line and decrease costs associated with the purchase of raw materials and with inventory control. Any resulting changes in operating efficiency or expenditures may help the company retain or improve its competitiveness in the marketplace.

Companies have traditionally evaluated their industrial processes in terms of production efficiency. Due to an increase in environmental concerns and changes in regulations associated with industrial waste, companies now incorporate waste management and prevention strategies into their industrial process evaluation. By precisely measuring and tracking waste generation and disposal, companies are provided with opportunities to evaluate production in new ways. Companies can greatly benefit from reexamination of their production lines with the goal of reducing waste generation. By increasing efficiency of operations, companies can direct more of the raw materials they buy into products rather than having the materials end up as waste.

Properly managing or treating waste for disposal is not considered pollution prevention. Continued focus on end-of-pipe technologies and treatments, while achieving regulatory compliance, really only perpetuates often inefficient manufacturing operations. Companies can improve efficiency by implementing a pollution prevention strategy that begins with the selection and purchase of materials before they arrive at the receiving dock. From a materials management point of view, pollution prevention is a strategy to use safer raw materials and to ensure their more efficient use.

Establishing and maintaining a pollution prevention program can be thought of as an eight-step process as shown in Figure 2.1. A pollution prevention program should be thought of as a continuous process, therefore the eight steps are arranged in a loop.

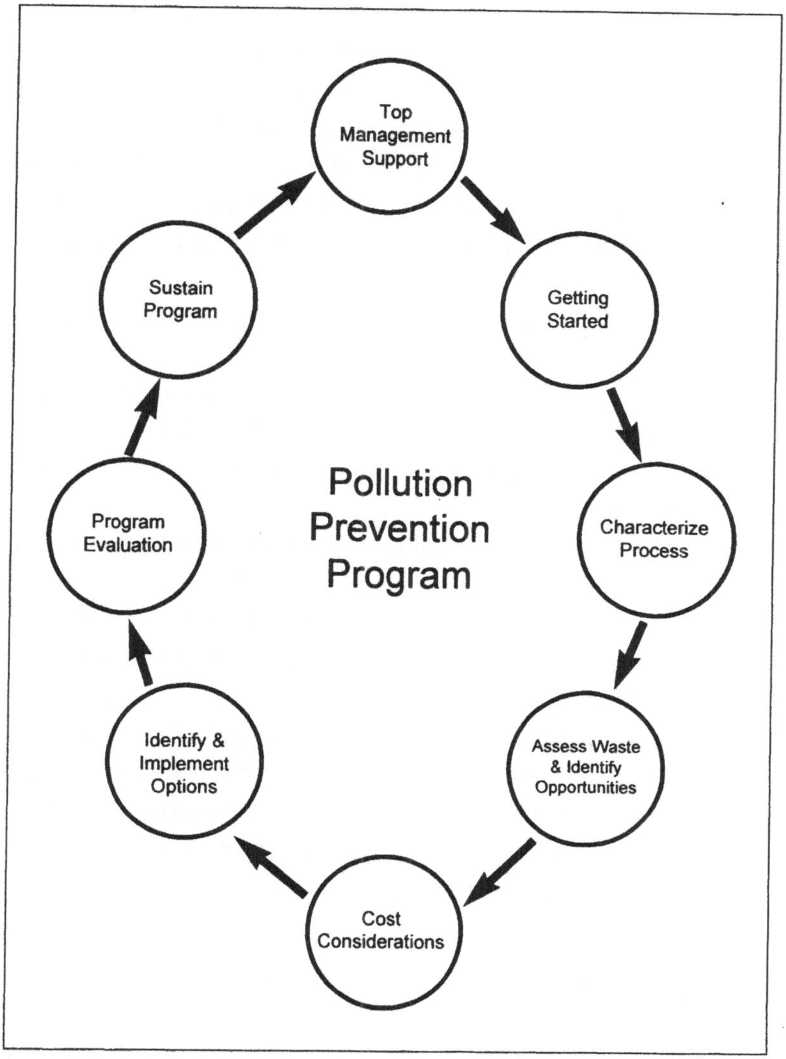

Figure 2.1. Pollution Prevention Loop. (*Source*: *Pollution Prevention: A Guide to Program Implementation*. TR-009. Champaign: Illinois Waste Management and Research Center, 1993.)

The eight steps that are outlined are as follows:

1. Obtaining support from top management
2. Getting the program started by beginning to incorporate the process within the company, developing a written pollution prevention plan, and training employees in pollution prevention
3. Reviewing and describing in detail the manufacturing processes within the facility to determine the sources of waste generation and to define a baseline inventory to be used to set goals and evaluate progress
4. Identifying potential pollution prevention opportunities for the facility
5. Determining cost of current waste generation and establishing a system of proportional waste management charges for those departments that generate waste
6. Selecting the best pollution prevention options for the company and implementing these choices
7. Evaluating the pollution prevention program on a companywide basis as well as evaluating specific pollution prevention projects
8. Maintaining and sustaining the pollution prevention program for continued growth and continued benefits to the company and reevaluating the program as economic situations change and process equipment requires upgrading.

Step 1: Top Management Support

Top management support is needed to get a pollution prevention program started, to incorporate it into already existing activities, and to sustain it. Getting support from the highest levels is important so that time and money can be allocated to practice pollution prevention. This step may prove to be the most difficult obstacle in establishing a successful pollution prevention program. Historically there are several approaches that result in management encouragement and participation in a pollution prevention program. Sometimes someone in management learns about such a program from outside the company and decides to implement one "at home." Frequently someone in the organization initiates a pollution prevention project on a small scale that demonstrates process efficiency or economic impact as well as environmental advantage. Such a demonstration can be used as a base on which to build an expanded activity. Often—particularly in smaller companies where the association between the production floor and top management is close—programs start by mutual decision, and top management support for a pollution prevention program is important. There are activities that can be planned and implemented in order to obtain such support. A pollution prevention program checklist is an important means by which top management may initiate and monitor the progress of implementation of a program. A typical checklist provides institutional, educational, technical, and economic steps.

Pollution Prevention Program Checklist

Top Management Support

- Write a policy statement supporting pollution prevention (see Figure 2.2)
- Distribute statement to all employees

Getting Your Program Started

- Commit to implementation

Sample 1: Management Policy Statement

(Adapted from Minnesota Office of Environmental Assistance, 1991)

We, [company name], are committed to excellence and leadership in protecting the environment. In keeping with this policy, our objective is to reduce waste generation and emissions. We strive to minimize adverse impact on the air, water, and land through excellence in pollution prevention. By successfully preventing pollution at its source, we can achieve cost savings, increase operational efficiencies, improve the quality of our products and services, and maintain a safe and healthy work place for our employees.

[Company name]'s environmental guidelines include the following:

- Environmental protection is everyone's responsibility. It is valued and displays commitment to [company name].
- Preventing pollution by reducing and eliminating the generation of waste and emissions at the source is a prime consideration in research, process design, and plant operations. [Company name] is committed to identifying and implementing pollution prevention opportunities through encouragement and involvement of all employees.
- Technologies and techniques which substitute non-hazardous materials and utilize other source reduction approaches will be given top priority in addressing all environmental issues.
- [Company name] seeks to demonstrate its corporate citizenship by adhering to all environmental regulations. We promote cooperation and coordination between industry, government, and the public toward the shared goal of preventing pollution at its source.

Sample 2: Management Policy Statement

At [company name], protecting the environment is a high priority. We pledge to eliminate or reduce, wherever possible: 1) our use of toxic substances, 2) our release of toxic pollutants, and 3) our generation of hazardous and other wastes.

When use of toxic substances, generation of wastes, or releases cannot be avoided, we are committed to minimizing any undesirable impacts on the air, water, and land.

Figure 2.2. Management Policy Statement. (*Source*: Minnesota Office of Environmental Assistance.)

- Designate a pollution prevention coordinator
- Set goals
- Increase employee awareness
- Train employees
- Recognize and reward pollution prevention successes

Understanding and Characterizing Processes and Wastes

- Gather background information
  1. Raw materials
  2. Waste generated
  3. Production mechanisms
  4. Process interrelationships
- Characterize general process
- Examine unit processes
- Perform materials balance
- Define production unit

Identify Pollution Prevention Opportunities

- Assess process
- Prioritize waste streams
- Generate reduction options

Cost Considerations

- Determine full cost of waste
- Develop economic charts
- Establish cost allocation system

Identifying and Implementing Pollution Prevention Projects

- Conduct technical evaluation
- Conduct economic evaluation
- Determine benefits

Program and Project Evaluation

- Modify program as needed
- Determine methods to measure waste reduction
- Evaluate program

Sustain the Pollution Prevention Program

- Rotate pollution prevention team
- Train employees as needed
- Publicize success stories
- Report regularly to management about program achievements and short- and long-term benefits to the company

To simplify the start-up, a new pollution prevention program might be incorporated into an already existing total quality management, environmental, or health program. To help win the support of high levels of management, an environmental manager may provide information on some of the benefits of implementing a pollution prevention program. An environmental manager should explain or demonstrate the potential for the following.

Reduced Operating Costs

An effective pollution prevention program often yields cost savings that more than offset program development and implementation costs. Cost reductions may be immediate savings that appear directly on the balance sheet, or anticipated savings based on avoiding potential future costs. Cost savings become more apparent when the costs resulting from the treatment, storage, or disposal of wastes are allocated to the production unit, product, or service that creates the waste. Achieving this may require broader approaches to accounting than typically have been used. However, assignment of costs to the product or process that generates them is a powerful tool to aid in the identification and prioritizing of pollution prevention opportunities. Material costs can be reduced by adopting production and packaging procedures that consume fewer resources, thereby creating less waste. When wastes are reduced, the percentage of raw materials converted to finished products increases, with a proportional decrease in material costs per unit of product. Waste management and disposal costs are an obvious and readily measured potential savings to be realized from pollution prevention. Federal and state regulations mandate special in-plant handling procedures and specific treatment and disposal methods for hazardous wastes. The costs of complying with these requirements and reporting on waste disposition are direct costs to businesses. Figure 2.3 illustrates how waste management costs have changed in the last thirty years, and the potential for saving money by reducing waste.

Production costs can be reduced through implementation of pollution prevention activities following an assessment. When a multidisciplinary group examines production processes from a fresh perspective, opportunities for changes that increase efficiency are likely to surface that might not otherwise have been noticed. Production scheduling, material handling, inventory control, and equipment maintenance are all areas that can be optimized to reduce the production of waste of all types and also control the costs of production. Often changes in the materials and operations of the process may be identified and implemented, resulting in significant economic advantage. Energy costs can decrease as pollution prevention measures are implemented in various production lines. There are also indirect costs, for example, higher taxes for public services such as landfill management. The expected trend is for these costs to continue to increase at the same or higher rates.

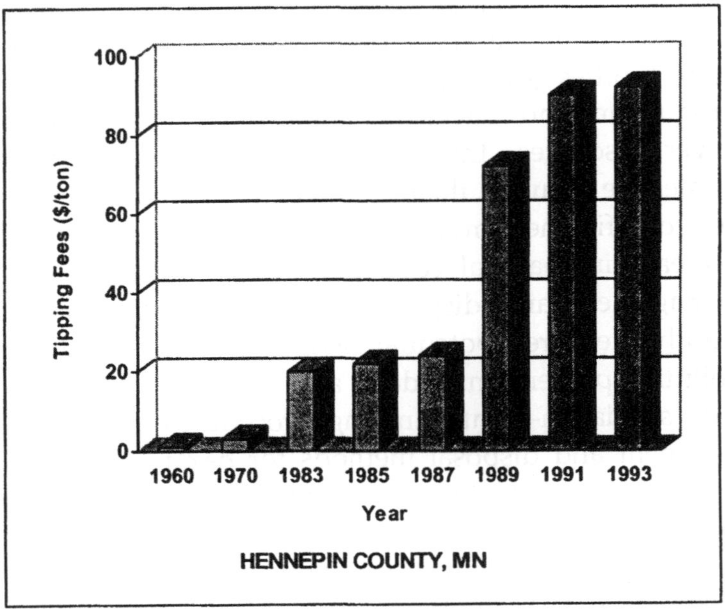

Figure 2.3. Waste Management Costs. (*Source*: Minnesota Office of Environmental Assistance).

Reduced risk of liability

A company may decrease its risk of both civil and criminal liability by reducing the volume and the potential toxicity of the vapor, liquid, and solid discharges it generates. A company should look at all types of waste, not just those that are currently defined as hazardous. Since toxicity definitions and regulations change, reducing the amount of wastes produced in all categories is a sound long-term management policy. Depending on location, environmental regulations at the state and local levels may require that facilities document the pollution prevention and recycling measures they employ for wastes defined as hazardous. Experience has shown that the potential for civil liability is increased by generating hazardous waste and other potential pollutants because of costs and impacts of later remediations of waste storage, treatment, disposal sites, and spills during transportation. Waste handling can affect public health and property values in the communities surrounding production and disposal sites. Even materials not currently covered by hazardous waste regulations may present a risk of civil litigation in the future. Workers' compensation costs and risks are often directly related to the volume of hazardous materials used and produced. It is unwise for a facility to confine its attention only to those materials specifically defined as hazardous.

Improved company image

As the quality of the environment becomes an issue of greater importance to society, a company's policy and practices for controlling waste and resource use generation increasingly affect the attitudes of employees and of the community at large. Employees may feel more positive toward their company when they believe that management is committed to providing a safe work environment and is acting as a responsible member of the community. By participating in pollution prevention activities, a positive interaction is established among management and employees. Helping to implement and maintain a pollution prevention program works to increase a sense of identity with company goals. This positive atmosphere helps to retain a competitive workforce and to attract high-quality new employees. Community attitudes will be more positive toward companies that operate and publicize a thorough pollution prevention program with some level of public participation, where appropriate. Some communities actively resist the siting of new waste management facilities in their areas. In addition, they are becoming more conscious of the monetary costs to society associated with treatment and disposal. Creating environmentally compatible products and avoiding excessive consumption and discharge of material and energy resources, rather than concentrating solely on treatment and disposal, can enhance a company's image within its community as well as with potential customers.

Step 2: Getting a Program Started

Although pollution prevention programs may start from several points, including individual smaller assessment and implementation projects, implementing a facilitywide program will probably require more organization and broader staff and management support. Before any assessments or evaluations are performed on plant activities as part of the facilitywide program, an environmental manager should work to win a commitment from all employees to implement a pollution prevention program. The time and effort involved will result in raised employee awareness, establishment of a cohesive pollution prevention team, and beginnings of incorporation of pollution prevention ideas into the day-to-day operations of the company.

A key element of a successful pollution prevention program is a capable pollution prevention coordinator. The coordinator has the responsibility of establishing the pollution prevention team, conducting meetings, and making sure the company is working toward its pollution prevention goals. More than likely, the coordinator will come from a midmanagement position. The coordinator will act as the key liaison to top management both to report progress and achievements and to secure resources.

With the coordination in place, it is time to establish a pollution prevention team. Members of the team should come from various departments within the facility, such

(Note: In small firms, the same person may wear several of these hats.)

Member	Responsibilities
1. Management	• Demonstrate corporate commitment. • Set and enforce long-term goals. • Have authority to implement changes.
2. Engineering and Design	• Provide information on current processes. • Contribute ideas for changing processes. • Gauge the technical feasibility of proposals. • Identify parameters for optimal functioning.
3. Environmental Compliance	• Calculate treatment and disposal costs. • Gauge environmental effects of proposals. • Ensure compliance with regulations.
4. Finance/ Purchasing	• Calculate costs of current operations. • Calculate costs and savings of proposals. • Implement changes in inventory control. • Track costs and benefits of actual changes.
5. Sales and Marketing	• Provide insight into customer needs. • Educate customers about TUR changes. • Market products as environmentally responsible. • Monitor customer reactions to product changes.
6. Production Workers	• Provide accurate descriptions of production practices. • Suggest ideas on new approaches. • Gauge compatibility of changes with work practices. • Supply feedback on front-line effects of changes. • Increase worker support for production line changes.

continued

Figure 2.4. The P2 Planning Team. (*Source*: The Office of Technical Assistance, Executive Office of Environmental Affairs, Commonwealth of Massachusetts, 1992.)

Larger companies might consider including:

Member	Responsibilities
Quality Control	• Gauge compatibility of changes with quality specifications.
Legal	• Gauge effects of TUR on regulatory requirements. • Gauge effects of TUR on corporate liability.
Safety and Health	• Provide data on health and safety effects of toxics. • Evaluate effects of TUR on worker health and safety.
Material Control and Inventory	• Provide information on material storage and use. • Suggest methods for improving storage and handling.
Research and Development	• Consider modifications to product design. • Consider TUR goals during new product development.

Figure 2.4. (continued)

as maintenance, production, environmental, health and safety, purchasing, shipping and receiving, legal and engineering departments, and plant and executive managers (see Figure 2.4).

A pollution prevention team will want to begin to formulate goals, even early in the pollution prevention program. A company might want to set an ultimate goal of zero waste generation to acknowledge the fact that pollution prevention is a continuing challenge. This is very similar to company goals like "zero product defects" or "zero lost workdays." Although this is a goal that may not be achievable, it can set a tone and degree of seriousness for the program that has long-term positive benefits. Another goal may be to replace some or all toxic substitutes used with nontoxic substances and thus reduce risk to employees, the public, and the environment. Numerical goals for waste reduction in contrast to percentage reduction may be established once the wastes are characterized. Ideas for strategies that may lead to goal achievement include incorporating pollution prevention progress into performance evaluations for all management staff, installing a revised accounting system that charges the cost back to the production line generating the waste, or training all employees in pollution prevention. Any goals or implementation strategies set at this stage of program formulation will, of course, change over time and become more clearly defined as employees work with them and attempt to achieve them.

To help everyone become aware of the pollution prevention program, communication throughout the plant is important. Supervisors should discuss the status of the pollution prevention program at regular intervals and encourage suggestions for improvement and additional progress. Mentioning the achievements of the pollution prevention program in a company newsletter is an excellent way to publicize the efforts. The following topic may be included in a written discussion of the start-up of a pollution prevention program:

- Provide a definition and explanation of the primary components of pollution prevention (i.e., source reduction, and in-process recycling).
- State company policy and guidelines clearly.
- Identify company goals to reduce waste generation and to improve operations.
- Stress the benefits of pollution prevention.
- Present facts on safety improvement that may occur when a pollution prevention program is implemented.
- Stress the relationship between the cost of generating waste and company competitiveness.

- Equate savings from pollution prevention with the company's fiscal health.
- Encourage employee participation in the effort to improve facility and environmental conditions.
- Make it known that management and pollution prevention team members are available and open to employee suggestions and new ideas.

To stimulate additional interest and continued participation in pollution prevention, an employee incentive award or recognition program for the facility or company may be established. Shifts, departments, and individuals can be encouraged to compete against their own past year's performance. The pollution prevention coordinator should look into applying for any awards given out by local, state, or federal environmental groups to facilities that practice pollution prevention successfully.

A facility coordinator should assemble a written pollution prevention plan. The written plan would include a statement from top management, an outline of goals and timetables, and a listing of those employees directly participating in the pollution prevention efforts. In some states, such plans are required for some companies. Experience has shown that for many companies, the preparation of a good plan can spark a vigorous and effective pollution prevention program.

Step 3: Understanding Processes and Wastes

Effective implementation of a pollution prevention program requires understanding of the various unit processes carried out at the facility and where in these processes waste is being produced. Large amounts of data gathering may be necessary to gain this understanding. General background information about processes and quantities can be found in Table 2.1.

Table 2.1. Possible Sources of Background Information.

Information on	From	
Raw Materials Use	Purchasing Records	Production Logs
	Inventory Records	Packaging Material Discarded
	MSDSs	Shipping and Receiving Logs
	Vendor Information	Annual Report
Waste Generated	Waste Manifests	Environmental Reporting
	TRI Data	Waste Collection and Storage
	Sewer Records	Production Logs
	Permits/Applicatons	Environmental Violations
	Flow Diagrams	Laboratory Analyses
	Annual Reports	Obsolete Expired Stock
	Rejected Product	Spill & Leak Reports
Production Mechanisms	Operations Manuals	Production Logs
	Vendor Information	Flow Diagrams
	Control Diagrams	Product Specifications
	Quality Control	
	Guidebook	

Table 2.1. (continued)

Information on	From	
Process Interrelationships	Product-to-Raw	Production Logs
	Material Data	Product Specifications
	Flow Diagrams	Facility Layout
	Quality Control Data	
	Dependencies on Preceding Processes	
Economic Information	Cost Accounting Reports	Pollution Control Costs
	Operating Costs for Waster Handling and Disposal	
	Costs for Products, Utilities, Raw Materials, Labor	

(*Source*: *Pollution Prevention: A Guide to Program Implementation*, TR-009. Champaign: Illinois Waste Management and Research Center, 1993.)

It is beneficial to define a production unit so that wastes generated can be quantified per production unit. This allows comparisons between the amounts of waste generated during different levels of production. A production unit can be individual items, pounds of materials, gallons of solution, or some other measurable quantity depending upon the nature of the products made. A typical process has raw material inputs, product outputs, and waste generation. This process can often be represented by a general process flow diagram. In addition to the raw materials, final products, and waste flows, other inputs can be represented on the general flow diagram, such as lubrication fluids, cleaning agents, or cooling water. This type of diagram will help provide an understanding of the relationships within the overall process and the associated wastes. Flow diagrams for unit processes can be created using either of two approaches:

1. Start with the wastes and products generated and then determine the sources of the waste by going backward through each of the unit processes.
2. Start with the raw materials and track them through each of the unit processes until products and waste material are generated.

An example of a process flow diagram for the metal parts fabrication industry is shown in Figure 2.5. Accuracy is critical when determining the types, compositions, and quantities of raw materials consumed, the product yield, and wastes generated for each unit pro-

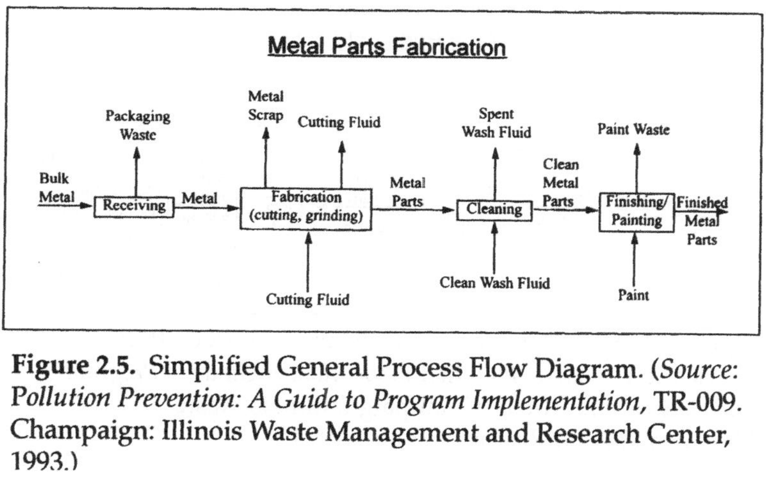

Figure 2.5. Simplified General Process Flow Diagram. (*Source: Pollution Prevention: A Guide to Program Implementation,* TR-009. Champaign: Illinois Waste Management and Research Center, 1993.)

cess. Different types of waste include emissions from stacks; vent emissions from process areas; fugitive emissions from pipes, tanks, or vessels and leaking equipment; spent washwater and cleaning solvents; cooling water; overspray from painting operations; cleaning rags; material scrap; and other wastes.

To fully understand a unit process, a materials balance can be performed (see Figure 2.6). A materials balance accounts for all inputs and outputs related to a process; in other words, what goes in must come out. A materials balance can indicate whether fugitive losses are occurring in the process and can determine how much raw material goes unconverted directly to waste. Once the materials balance has been performed, the actual amount of each waste and its sources become apparent. These numbers are the baseline amounts of total waste generation at the start of the pollution prevention assessment and can be used for comparison throughout the implementation of the program.

Once materials balances have been made, a waste stream summary sheet similar to Table 2.2 can be filled out for every waste stream. Arranged in this fashion, dif-

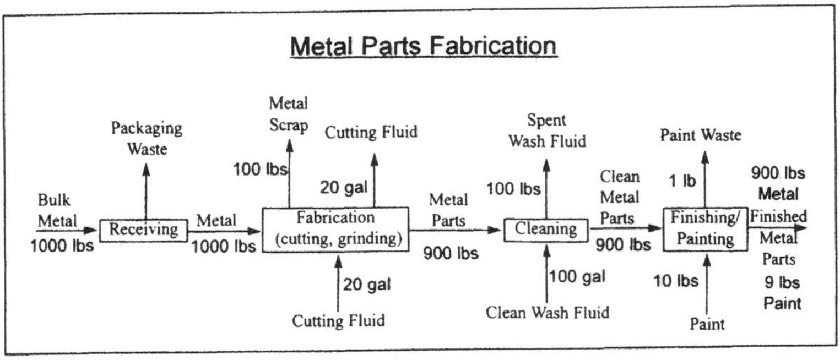

Figure 2.6. Simplified Materials Balance for Metal Parts Fabrication. (*Source*: *Pollution Prevention: A Guide to Program Implementation*, TR-009. Champaign: Illinois Waste Management and Research Center, 1993.)

Table 2.2. Waste Assessment Survey Form. (*Source*: Elizabeth Treadway, Environmental Services Director, City of Greensboro, NC.)

Date:________________
Facility Location:________________
Persons Interviewed:________________
(Title and/or Name)

WASTE STREAM	
Content	
Quantity Generated	
Frequency of Generation (seasonal changes, emergency vs normal operations)	
Generating Process	
Original Material Source/Supplier/Cost	
Current Disposal Costs	
Current Handling/Disposal Mtds (written procedures?)	
Current Waste Min. Practices (change in volume generated, written procedures?)	
Proposed/Planned/Suggested Waste Minimization Procedures (employee suggestions)	
Flow Diagram	

ferent waste streams can be compared to learn which waste streams are the most environmentally harmful and costly to a facility. Table 2.3 can help show if waste streams are mixing at a facility before being discharged. Segregation of different waste streams can lead to sub-

Table 2.3. Waste Handling Procedures and Resources. (*Source*: Elizabeth Treadway, Environmental Services Director, City of Greensboro, NC.)

Is waste handled in separate streams?

Indicate which waste streams are generated at the facility, and show any grouping of streams by current handling practices:

_____ Office	_____ Public Areas	_____ Conference/Meeting Rooms
_____ Kitchen	_____ Eating Areas	_____ Snackbar areas
_____ Research Areas	_____ Restrooms	_____ Experimental Areas
_____ Manufacturing	_____ Machining	_____ Chemical Labs
_____ Shipping	_____ Receiving	_____ Material Storage/Distribution

How many hours of custodial time is spent per week handling waste?

____

What percentage of overall custodial time does this represent?

____

What waste, if any, is not handled by custodial staff?

____

Other comments: ______________________________

How waste is collected, handled and transported for disposal?

(Example: Offices--Once a night, custodians pull waste liners from individual waste baskets, put bags in 44 gal. barrel, and take barrel to waste dumpster at end of shift.)

Problems, obstacles and challenges to handling waste within the building:

Other comments:

---

stantial cost savings from treatment or disposal costs, as well as increased opportunities for recycling or reuse. Table 2.4 is an itemized breakdown of the different types of waste that may emerge from different departments of a facility.

Step 4: Assessing Wastes and Identifying Pollution Prevention Opportunities

Waste streams and unit processes should be prioritized based on cost or environmental impact in order to determine the best order of examination leading to implementation of pollution prevention opportunities. The following factors can be considered when ranking the streams:

- High purchase and/or disposal cost
- Highly toxic waste
- Waste regulated under the Toxic Release Inventory (TRI)
- Waste regulated under Resource Conservation and Recovery Act (RCRA)
- Particular regulatory concerns
- High use or release rate
- Potential liability due to endangerment of employees, environment, or the public
- Carcinogens
- Hazardous Air Pollutants (HAPs)
- Chlorofluorocarbons (CFCs) and other materials that might be banned in the future
- Local citizens' concerns

Once high-priority streams have been targeted, the flow charts developed in the previous steps provide insight into potential control points and may indicate

Table 2.4. Waste Composition. (*Source*: Elizabeth Treadway, Environmental Services Director, City of Greensboro, NC.)

		Offices	Production	Eating Areas	Other	Total
% of Bulk CY per Week		____	____	____	____	100%
Paper						
Newspaper		____	____	____	____	____
Books & Magazines		____	____	____	____	____
Office Paper	White, CPO	____	____	____	____	____
	Colored	____	____	____	____	____
	File stock	____	____	____	____	____
	Mixed	____	____	____	____	____
Cardboard	Corrugated	____	____	____	____	____
	Paperboard	____	____	____	____	____
	Waxed	____	____	____	____	____
Other Paper	Tissue, Wrap	____	____	____	____	____
	Glossy	____	____	____	____	____
	Other	____	____	____	____	____
Compostables						
Wood	Brush, Trees	____	____	____	____	____
	Pallets	____	____	____	____	____
Yard Waste		____	____	____	____	____
Food		____	____	____	____	____
Soiled paper		____	____	____	____	____
Metals						
Aluminum		____	____	____	____	____
Ferrous Metal		____	____	____	____	____
Other Non-Ferrous		____	____	____	____	____
Other Materials						
Textiles	Organic	____	____	____	____	____
	Non-Organic	____	____	____	____	____
Glass	Clear	____	____	____	____	____
	Green	____	____	____	____	____
	Brown	____	____	____	____	____
Plastics, Rigid	HDPE	____	____	____	____	____
	LDPE	____	____	____	____	____
	Styrene	____	____	____	____	____
	Other	____	____	____	____	____
Plastics, Film	HDPE	____	____	____	____	____
	LDPE	____	____	____	____	____
	Styrene	____	____	____	____	____
	Other	____	____	____	____	____
Miscellaneous		____	____	____	____	____
Other Materials						
Stone, Clay, Sand		____	____	____	____	____
Sludge		____	____	____	____	____
Total		100%	100%	100%	100%	100%
Other Comments						

directions for a more detailed investigation. The process will need to be studied in the facility, both while the process is in operation (ideally during all shifts) and, if possible, during a shut-down/clean-out/start-up period to identify what materials are used and what wastes are generated by this procedure. While studying the process, one should note any potential pollution prevention opportunities, paying attention to the following:

- Procedures of operation by line workers including material handling and transfer operations
- Quantities and concentrations of materials (especially wastes)
- Collection (including exact sources) and handling of waste (note if wastes are mixed)
- Any record keeping including records of spoiled batches and their disposition (obtain copies of these if not already done)
- Flow diagrams (follow through actual process)
- Leaking lines or poorly operating equipment
- Any spill residue
- Damaged containers
- Physical and chemical characteristics of the wastes and releases

The pollution prevention team (listed in chart form in Figure 2.7. as the Waste Reduction Audit Team) should determine the optimum operating conditions, product specifications, and equipment specifications for the process. Specifically, the team should try to identify when waste is generated—not just by the regular process but by upsets, off-specification products, and spills, for example. The team should also talk with the maintenance and housekeeping personnel (including outside contractors where appropriate) who service the process to determine when, why, and how the process is maintained.

Listed in the order of ease of implementation, benefit from result, and cost of implementation, some common waste reduction alternatives, include:

- Housekeeping and inventory controls

The Waste Reduction Audit Team

Company: ____________________

Date: ____________________

All employees with questions or suggestions regarding waste management and waste reduction are encouraged to contact any one of the people listed below:

Team Leader: ____________________

Title: ____________________

Team Leader: ____________________

Title: ____________________

Team Leader: ____________________

Title: ____________________

Team Leader: ____________________

Title: ____________________

Audit Team Goal: ____________________

____

____

____

Please Post

Figure 2.7. The Waste Reduction Audit Team. (*Source*: David Wigglesworth. *Profiting from Waste Reduction in Your Small Business*. Anchorage: Alaska Health Project, 1988.)

- Substituting a nonhazardous material for a hazardous one
- Substituting a material that reduces the volume or toxicity of the wastes the process generates
- Process elimination

Housekeeping and Inventory Controls

Housekeeping and inventory controls are some of the easiest hazardous waste reduction alternatives to implement, as well as the least expensive. These alternatives can be as simple as changing the way you handle waste or the way you order materials. Implementing these controls may include training your personnel in better handling and procurement procedures.

The following series of questions is aimed at identifying possible housekeeping and inventory improvements within your facility.

Do you have designated areas where wastes are accumulated? Keep accumulation areas clean and uncluttered. Remember, all hazardous waste accumulation containers should be kept covered at all times. Label containers depicting which wastes should be accumulated there, so that you can avoid mixing different waste types, and, more importantly, you can avoid mixing nonhazardous with hazardous wastes. Train employees how to use these containers and in the importance of keeping different wastes separated. For example:

- Do not mix oily waste with solvent-containing wastes.
- Do not discard solvent-laden rags used to clean parts with rags used for other purposes.
- Do not throw out ordinary trash, such as sandwich wrappers, in hazardous waste accumulation containers.

- Do not mix absorbants used to collect oil spills and leaks with regular oily wastes (especially if your waste oil is sent offsite for recycling).
- Do not accumulate used oils in empty solvent or pesticide containers; all containers should correctly identify their contents.

Do you store hazardous materials in designated areas? Keep your hazardous materials stored in designated areas. When not using hazardous materials, return them to the storage area. Keep your material storage area and your waste storage area separate. Have Material Safety Data Sheets (MSDSs) available near the point of storage for all the hazardous materials you use.

Do you completely use your hazardous materials before discarding them? Use your hazardous materials according to manufacturer's instructions. You do not have to discard hazardous materials until they can no longer be used for their intended purpose. For example:

- When using solvents and rags for cleaning, apply solvent to rag just prior to use, and keep the solvent container covered.
- Do not discard rags until you can no longer effectively use them.
- Do not discard degreasing solvents until they no longer effectively clean.
- Use one container of material at a time.
- Completely use hazardous materials before discarding the container.
- Change motor oil and oil filters only when necessary.

Are you currently storing materials that you do not use? Inspect all the material storage areas to see if material is present that is no longer used. You should make it standard procedure in your department to store only those materials that are used in your operations. Materials past their expiration date are considered wastes and must be properly discarded. Consider finding a market, specific business, or another department in your company that may be able to use your unwanted materials. Some areas have waste exchanges to broker transactions.

Do you order more material than your department can use? During your inspection, check to see if materials are expiring before being used. It may initially appear to be cost effective to order materials in bulk. However, when you factor in the cost of discarding the material you were unable to use before the expiration date, buying in bulk may not seem so favorable.

The advantages of housekeeping and inventory controls as a hazardous waste reduction alternative include:

- Easy to implement
- Cost effective
- Increased worker safety
- Less material stored in your department
- Chance of accidental spills or leaks is decreased.
- Reducing the amount of material used generates smaller quantities of wastes, so less waste will need to be handled, transported, and disposed.
- Need to transport less hazardous material to your facility
- Easier preparation of the Hazardous Waste Manifest when accumulation areas are labeled and uncluttered
- Chance of being fined for improper management of hazardous waste is reduced.

Typical housekeeping and inventory controls include:

- Purchase covered containers for the accumulation of oily rags and rags with solvents.
- Provide hazardous materials and hazardous waste management training for employees.
- Prepare standard labels to be used on hazardous waste accumulation containers. This should be an easy task for those wastes you routinely generate.
- Inspect your department on a regular basis (weekly) to assure that housekeeping controls are being followed and general housekeeping is improving.
- Keep materials stored in locked areas. Assign one person the responsibility for assuring new materials are distributed only when previously distributed materials are spent. Some facilities have been initiating a "pharmacy" approach for dispensing where only the amount of materials required for a specific job is given to the worker.
- Post signs instructing employees on proper housekeeping procedures (e.g., CONTAINER MUST BE COVERED).
- Order only those materials used by the department.
- Order materials in quantities that can be fully used before the expiration date.
- Adopt the FIFO (first in—first out) rule while using materials in stock.
- Dispose of all materials currently stored that are not being used. If the material has not expired, try to give it to another department that uses this material instead of discarding it. Many universities or other schools will take nonexpired chemicals for their laboratories, although they should do this selectively to assure that they do not acquire material they will pay to dispose of later.
- Track the amount of materials used and adjust ordering accordingly.

Substituting a Nonhazardous or Less Hazardous Material for a Hazardous Material

If practical substitutions exist, this alternative can be very easy to implement. To be successful, substitutions of nonhazardous or less hazardous materials for hazardous ones should be coordinated through your facility's purchasing group. This may include developing a list of suitable nonhazardous or less hazardous materials that may be purchased, evaluating each purchase requisition to see if nonhazardous or less hazardous alternatives exist, and restricting the purchase of hazardous materials if suitable nonhazardous or less hazardous materials do exist.

For this hazardous waste reduction alternative to prove practical at your facility, the material that is acting as a substitute should provide the same function as the material being substituted. The following questions need to be asked:

1. Will the substitute work as well?
2. Will the substitute generate a different type of waste stream or release?
3. Will the substitute require purchasing additional equipment?
4. Will the substitute cost more?
5. Will the substitute result in releases to the air, water, or land?
6. Will the substitute violate any maintenance specifications?

Onsite Recycling

The objective behind onsite recycling is to take a waste material and recycle it onsite so that it can be used as a raw material in a facility's processes. Although this is potentially an environmentally positive operation, it is not source reduction unless "in-process" recycling is carried out. "In-process" recycling is essentially defined as a recycling operation hard-piped into the process. In other words, the hazardous material never leaves the process itself through accumulation in storage containers, for example. Onsite recycling typically involves some investment in capital equipment and additional personnel, which makes it more difficult to implement than the previous alternatives considered. This investment may prove worthwhile at your facility depending on the amount of wastes that can be recycled. Another consideration is the centralization of the onsite recycling. In other words, one maintenance shop may not generate enough used oil to make the onsite recycling economically feasible. However, if used oil from all of the maintenance shops and vehicle repair areas at a facility were brought to a centralized onsite recycling unit, this option may prove feasible. Oil, solvents, and empty containers have been successfully recycled onsite.

Advantages

- Reduced liability since the waste is not being transported offsite
- Elimination of manifesting requirements
- Reduction of raw material costs
- Reduction of disposal costs since waste is not being disposed of offsite

Other considerations

- Frequently, the recycling technology itself will result in a wastestream. Solvent distillation residues, for example, most often will have to be managed as hazardous.
- Capital and operating costs may be incurred for the recycling equipment.

- State and federal hazardous waste facility permits may have to be obtained, since your facility will be treating the waste onsite (depends on the waste stream and the recycling method).
- Increased health and safety concerns may result for employees who will have to operate equipment.
- Hazardous waste may be transformed from one form to another.
- Potential for releases to other media may be created.

Offsite Recycling

Transporting a waste material to an offsite recycling facility does not typically involve any investment in capital equipment and additional personnel, and is not a source reduction technique. Some arrangements may include "roundtripping," in other words, the recycling facility returns the recycled material back to the generator for use.

Examples of some situations where offsite recycling has been successful include oil, solvent, and empty container recycling and the laundering of oil- and solvent-contaminated rags.

Advantages

- Raw material costs may be reduced if recycled materials are returned to the facility for use in its processes.
- Disposal costs are reduced. The cost for offsite recycling varies with market conditions. However, the costs are often not more than the transportation costs. In some instances the recycler may pay the generator for the waste.
- There is less liability than when disposing of the waste at a hazardous waste facility.

Other considerations

- Compared with onsite alternatives, liability may increase, since the waste is being transported offsite.
- Manifesting is required.
- Hazardous waste may be transformed from one form to another.
- Potential for releases to other media may be created.

Process Elimination

The objective behind this hazardous waste reduction alternative is to eliminate unnecessary processes that generate hazardous waste. This will probably be the most difficult alternative to implement since most of a facility's processes are probably necessary to daily operation. However, it may prove worthwhile to think about some routine operations where the frequency may be changed. For example, does your facility change the oil in all pieces of equipment at the same time even if an oil change may be unnecessary for some equipment? Is it necessary to clean according to the current frequency or can you reduce how often you perform general facility cleaning? Can small quantities of leftover paints of different colors be mixed and used for painting new surfaces? By thinking creatively, regular procedures—especially cleaning procedures—can be either modified or eliminated to minimize hazardous waste generation.

Advantages

- The waste stream is eliminated entirely.
- This reduces liability, improves worker health and safety, and prevents releases to other media.

Other considerations

- Costs for replacements may result.

- Implementation may be difficult.

Step 5: Cost Considerations and Pollution Prevention Finance

With the change in environmental philosophy from end-of-pipe waste treatment to pollution prevention at the source, the cost accounting of potential environmental projects has become more difficult. Pollution prevention occurring upstream in an industrial process leads to a number of repercussions midstream and downstream. Changing material inputs from more- to less-toxic substances, for example, necessitates careful consideration of how such changes will affect equipment performance and product quality. Similarly, product redesign may require equipment modifications within a process line. Accompanying these technology adjustments will be a set of capital and operating costs which may be far-reaching, indirect, and not immediately evident. Thus, when evaluating the profitability of pollution prevention investments, compiling and analyzing operational and capital costs becomes that much more complicated and prone to error. Accurate costing quickly moves beyond the capabilities and data sources of a single staff person. Contributions from multiple departments such as the environmental engineering, the materials management, the production engineer, and the finance will be necessary to assemble such data. Table 2.5 gives a thorough listing of many different, often "hidden," costs to consider when determining the full cost of a waste stream. Table 2.6 is one attempt to quantify waste costs from many different departments at a facility.

A second reason why pollution prevention projects are difficult to cost is that they are associated with contingent benefits, which are difficult to predict and quantify.

Table 2.5. Costs to Consider—Determining Full Costs of a Waste Stream.

Hazardous Substance Use	**Waste Generation**
Purchasing	**Air and Water Emissions**
Taxes on hazardous products	Air emission permits and controls
Safety training	TRI measurements/estimates
MSDS filing	TRI reporting
Safety equipment	TRI fees
Extra insurance permiums	Worker health monitoring
Labor	Sewer discharge fees
	NPDES permits
	Water quality monitoring
Storage and Inventory	Sampling training
Special storage facilities	Pretreatment equipment
Safety equipment	Pretreatment system operation
Storage area inspection and monitoring	
Storage container labeling	**Solid Waste Collection**
Safety training	Safety training
Emergency response planning	Safety equipment
Spill containment equipment	Collection supplies
Lost product from spills, evaporation, etc.	Container labels
Labor	Container labeling
SARA Title III (TRI) reporting	Recordkeeping
	Truck maintenance (for in-house fleet)
In-Process Use	**Waste Storage**
Safety training	Storage permits
Safety equipment	Special storage facilities
Containment facilities and equipment	Spill containment equipment
Clean-up supplies	Emergency response planning
Labor	Safety training
	Storage area inspection and monitoring

continued

Table 2.5. (continued)

Hazardous Substance Use	Waste Generation
Lost Raw Materials	**On-Site Treatment or Recycling**
Labor for handling	Capital and operating costs
Equipment for clean-up	Depreciation
Reporting	Utilities
	Operator Training
	Safety equipment
	Emergency response planning
	Permits
	Inspection and monitoring
	Insurance
	Disposal
	Sewer fees
	Container manifesting
	Disposal vendor fees
	Preparation for transportation
	Transportation
	Insurance and liability
	Disposal site monitoring

Avoided liability is the prime example of this situation. The firm examining a potential project aimed at eliminating a hazardous wastestream or emission may be motivated by a desire to eliminate certain risks of litigation linked to personal or property damages. However, such risks are probabilistic in nature; that is, they may materialize if and when an accident occurs or a claimant sues the firm for such damages. Introducing this future monetary benefit into a project profitability analysis is problematic. The benefits depend on whether, when, and how much liability cost is avoided.

Third, profits from prevention projects may materialize well beyond the two- to five-year time frame commonly applied in investment analysis. Though the

Table 2.6. Business Waste Reduction Assessment Worksheet (*Source*: Elizabeth Treadway, Environmental Services Director, City of Greensboro, NC.)

Part 1. Facility Contacts

Facility Name:____________ Date:____________ Assessor:____________
Address:____________________________________
Main Contact:____________ Title:____________ Tel. No. ____________
Chief Exec. Officer:______________________ Tel. No.____________
Building Manager:______________________ Tel. No.____________
Custodial Supervisor:________ ____________ Tel. No.____________
Salvage Supervisor:______________________ Tel. No.____________
Material Handling Supervisor:______________ Tel. No.____________
Purchasing Agent:______________________ Tel. No.____________
Other Contact:____________ Title:____________ Tel. No.____________

Part 2. Business and Facility Types

Major Products Manufactured or
Produced:______________________________
Annual Sales:____________ No. of Employees:____________ Year Established_____
Company Purpose or Assessment: ____________________

Business Type	% Space	# Empl.	Business Type	% Space	# Empl.
___ Office	_____	_____	____ Retail	_____	_____
___ Manufacturing	_____	_____	____ Shipping	_____	_____
___ R&D	_____	_____	____ Warehouse	_____	_____
___ Other	_____	_____	____ Other	_____	_____

Part 3. Waste Volume, Hauling Method, and Cost Estimates

Dumpster, Volume, & Cost Characteristics

No.	Size	Type	Compacted?	Service/WK	% Full	Wks/Yr	CY/Yr
___	___	___	______	______	_____	_____	_____
___	___	___	______	______	_____	_____	_____
___	___	___	______	______	_____	_____	_____

Hauler(s):________________ Hauler Confirmation?______________
Est. Tons per Year____________ Est. Cost per Year______________

Other Comments:

---

savings of some practices—such as improved inventory control—may accrue in the short term, major and more costly changes to processes, materials, and products are likely to extend well beyond the two- to five-year time period. Analytical methods that fail to capture future

Table 2.7. Operating Costs. (*Source: P2/Finance, Version 2.1.* Boston: Tellus Institute, 1995.)

Current Process			**Alternative Process**		
Item	**Annual Cost ($/yr)**	**Total**	**Annual Cost ($/yr)**	**Total**	**Diff. = Curr. - Alt.**
Direct Materials					
Waste Mgmt. (materials and labor)					
Pre-treatment					
On-site Handling					
Storage					
Treatment					
Hauling					
Insurance					
Disposal					
Utilities					
Electricity					
Steam					
Water					
Sewerage					
Direct Labor					
Regulatory Compliance (materials and labor)					
Insurance					
Revenues-sale of Product					
Revenues-Marketable By-Products					
TOTAL					

streams of savings contain inherent biases against clean technology investments. Table 2.7 and Table 2.8 attempt to deal with evaluating costs over a multiyear time frame. (Table 2.7. lists operating costs associated with a modified process, ranging over many years. Table 2.8 lists capital costs, thought to be one-time expenditures.)

Clearly, the current full cost of waste generation at a facility must be determined before any pollution prevention project can be properly considered. Once this is

Table 2.8. Capital Costs for Proposed Project. *(Source: P2/ Finance, Version 2.1.* Boston: Tellus Institute, 1995.)

Capital Costs	**Cost**	**Totals**
Purchased Equipment		
Delivery		
Sales Tax		
Initial Spare Parts		
Materials		
Piping		
Electrical		
Instruments		
Structural		
Insulation		
Utility Connections and New Utility System		
Electricity		
Steam		
Water		
Sewerage		
Refrigeration		
Fuel		
Plant Air		
Inert Gases		
Site Preparation		
Demolition and Cleaning		
Old Equipment/Rubbish Disposal		
Construction/Installation		
In-house		
Contractor		
Vendor		
Engineering/Contractor		
In-house Planning		
In-house Engineering		
Procurement		
Contractor/Consultant		
Start-up/Training		
In-house		
Vendor/Contractor		

continued

Table 2.8. (continued)

Trails/Manufacturing Variances		
Training		
Contingency		
Permitting		
Fees		
Contractor/Consultant		
Initial Charge for Catalyst and Chemicals		
Working Capital		
Raw Materials		
Materials and Supplies		
Product Inventory		
Salvage Value		

done, the economic feasibility of pollution prevention techniques, including calculating the cost savings and payback periods, can be determined.

Regardless of whether you are performing a cost analysis by hand or by computer, several considerations are key to analytical effectiveness. In this section, these considerations will be discussed along with a number of analytical methods that can be used to measure the relative economic benefit of an investment.

One of the simplest measures of profitability, which does not consider the time value of money, is the *Return on Investment* (ROI). The ROI is a measurement of investment performance, calculated as the ratio of annual net income (gross income minus depreciation) over the initial investment amount. The equation can be written as:

$$\%\ \mathrm{ROI} = \frac{\text{Annual net income}}{\text{Initial investment amount}} \times 100$$

Payout time, or the time in years it takes to recover funds invested in a project, is another simple method that can be used to evaluate economic feasibility. The equation is written as:

$$\text{Payout time (yr)} = \frac{\text{(Fixed capital + Start-up costs) (\$)}}{\text{(Profit after taxes + Depreciation)(\$ / yr)}}$$

The variables are defined as follows:

Fixed capital: the costs required for process construction
Start-up costs: costs incurred from process modifications and start-up labor, which also include revenue loss during process debugging
Profit after taxes = (Annual Revenue – Annual Product Cost)
Depreciation: the reduction in the equipment's initial value over time, measured in $/year. Depreciation involves expensing the equipment over its life span, as an offset to earnings. The IRS allows a company to recover an annual amount, by means of a reduction in tax liability, that is representative of the yearly reduction in the equipment's initial value. Straight-line depreciation involves deducting 33% per year of the value of equipment having a 3-yr life, 20% per year for equipment having a 5-yr life, 10% for equipment having a 10-yr life, and so on.

In order to increase accuracy in estimating a project's potential worth, one can consider the time value of money, which enables one to take capital cost (generally a single expenditure of money) and put it on the same basis as dollars per year—the common unit upon which operating costs are based.

The *discounted cash flow analysis* takes into account the time value of money. This technique allows for translating future cash flows into values indicative of the same period within which the initial capital investment was made. One popular financial indicator that incorporates the discounted cash flow analysis is *Net Present Value* (NPV).

In the NPV calculation, the value of cash inflows and outflows are equalized into "present" values, the sum of which is the project's NPV. Before getting into an illustrative example demonstrating the use of NPV, it is important to go over some general rules of thumb with regard to NPV.

- If NPV = 0: The project generates exactly the return that is required; the amount saved by the project is sufficient to pay off the initial outlay of funds, pay off interest payments to creditors who lent money and provide the required return to shareholders.
- If NPV > 0: The project should be accepted; the amount saved by the project is sufficient to accomplish all of the above plus increase the economic value of the business.
- If NPV < 0: The project should be rejected.

Example

An electroplating company is considering recycling process acid by diffusion dialysis. After substantial research, the facility has determined the following numbers:

- Costs:
  1. Equipment costs (one-time): $8,000
  2. Installation (one-time): $1,000
  3. Annual Operating Costs: $2,000
- Savings: The project will generate $6,000 in savings in each of the next three years.
- Therefore, the net annual cash flow is $6,000 – $2,000 = $4,000.

Assuming a discount rate of 14%, should the project be implemented?

Solution:
NPV at a 14% discount rate

Year	Cash Flow	Discount Factor	Present V.
0	–$9,000 ×	1.000 =	–$9,000
1	$4,000 ×	.8772 =	$3,509
2	$4,000 ×	.7695 =	$3,078
3	$4,000 ×	.6750 =	$2,700

Net Present Value of Cash Flows over 3-yr period = $287
NPV > 0, good investment
IRR = 15.9%

Considering a continuous interest rate, the discount factor is calculated with the formula:

$$1/(1+i_c)^n$$

i_c = continuous interest rate
n = the number of years

To further refine your approach in screening potential projects, you may wish to organize capital, operating, and future liability costs in such a way that relevant data are easily identified. Keep in mind that capital costs occur once for a project and include equipment, construction materials, installation, engineering, start-up, training, and permitting. In addition, remember to take advantage of tax breaks by incorporating a depreciation schedule. The real challenge lies in collecting the data. Challenges include finding the right sources of information and identifying who did the work. For example, you may need information on manifesting time (environmental engineer), wage rate (payroll department), and quantities of raw materials required for particular processes. With good internal communication, a facility can effectively price different pollution prevention projects, and make decisions that benefit both the environment and the bottom line.

Step 6: Identifying and Implementing Pollution Prevention Projects

Once suggestions for pollution prevention options are gathered and the costs and cost benefits associated with these options calculated, they should be reviewed by the pollution prevention team, who should eliminate the least beneficial options from further consideration. These options may be reviewed again at a later time, since pollution prevention is a continuous process. What is less beneficial now. may work better in the future. The remaining options should then be examined in more detail to determine overall benefits. Technical and economic feasibility of each option, based on the company's requirements for these criteria, should be studied. For example, each company has its own standards for economic evaluation, feasibility, levels of expertise, and operational requirements. Those options found to be consistent with the company's goals can then be scheduled for implementation. There may even be cases in which certain benefits of a project override low economic return.

In implementing a P2 modification, it is critical to coordinate the efforts of all departments and personnel involved in the change. Many P2 modifications occur upstream in a process, and have impacts on many different departments. It is also likely that some special training may be necessary for workers to handle any new machinery. Successful implementation does not end when new equipment is installed, and initial worker training is completed. Inevitably there will be some unpredictable snags and difficulties. It is critical that the P2 team study the changes to make sure they are working as planned, and make modifications to procedures until all effected parties are satisfied.

Finally, the P2 team should make sure that they have a good understanding of the waste streams that are the result of the *existing* process that is to be modified. Of course, this needs to be understood so that any cost savings can be calculated. Once modifications are made, it is too late to measure previously existing waste streams.

Step 7: Program and Project Evaluation

After a period of time, the P2 team will want to measure progress against its goals. By reviewing the program, managers at all levels can assess the degree to which pollution prevention goals and production targets at the facility are being met and what the economic results have been. The comparison identifies pollution prevention techniques that work well and those that do not. The information will help guide future pollution prevention assessment and implementation cycles.

Quantitative evaluation also enables managers to compare their units with similar units in their company and with data from other companies. The P2 team will need this benchmarking knowledge to plan enhancements of the current pollution prevention program, to select technologies for transfer from other operations, and to help identify new pollution prevention options.

Acquiring Data

The employee charged with monitoring the success of a pollution prevention project needs to determine a measurable indicator of waste management performance (e.g., waste volume or toxicity), measure that quantity, and normalize the data, as necessary, to correct for factors not related to the pollution prevention method being reviewed. For example, waste generation per unit of production is one normalization approach. Although the process is simple in theory, complexities arise in practice. There are a number of factors to consider when defining what data you will track.

First, the indicator selected to track performance must accurately reflect the waste(s) of interest. Second, the indicator must be measurable with the resources available to you. Material and energy balances will be helpful in organizing data and can help fill in some gaps in data.

After deciding what data should be tracked, you will need to determine how to collect it and what normalization may be required for each category of data. Below are some considerations when evaluating the success of a pollution prevention project.

Regulatory Reporting Data

Depending on the type of business your company engages in, you may have a considerable volume of data already collected for regulated waste streams. However, there can be gaps and discrepancies in this data. For example, RCRA wastes are characterized by waste type and total amount, but not by individual components. Therefore, this data may not be specific enough for your evaluation. In addition, accurate measuring devices may not be available for all waste streams (e.g., vaporous or fugitive emissions). In such cases, your regulatory compliance records would have been based on estimates; comparing estimates from one period to another will not yield very reliable percent-of-change figures. Finally, year-to-year comparisons may not be meaningful if the reporting requirements changed sufficiently to cause differences in how waste quantities were measured.

Waste Shifted to Other Media

Pollution prevention may eliminate part of the target material but shift some of it to another plant stream, to another environmental medium, or into the product. It can be difficult to track the shift of a pollutant from one medium to another or to determine what new pollutants may be created by the new procedure. Conducting a mass balance analysis would help. Transferring a given pollutant to another medium or replacing it with a different pollutant should be avoided, unless there is a clear environmental benefit such as decrease quantity or toxicity of the waste. If you were to find that transfer was occurring, you would need to evaluate very carefully the relative impact on the environment.

Measuring Toxicity

The toxicity of the waste should be considered, as well as the quantity produced. Reducing the sheer volume of a given waste product while increasing its per-unit toxicity is a treatment option, but it is not pollution prevention. For example, adding lime to a waste stream to precipitate metals reduces the volume of waste but does not prevent pollution since the total quantity of metal is not changed. Since toxicity frequently is not measured as part of production reporting, you may have to establish procedures for doing so. Comparison of relative toxicities or risks from various chemicals is not an easy task because the necessary information is not always readily available. Often supplies can provide relative risk information about a small number of options. Material Safety Data Sheets may also be used to compare risks.

Normalizing for External Factors

Changes in quantity are straightforward, easily understood, and relatively easy to calculate if data are available. Quantity comparisons from one period to the next can be useful input to a pollution prevention program review. However, the data will have to be normalized if there were major factors unrelated to pollution prevention efforts that influenced the quantities produced.

A number of external factors can cause the quantities or mix of products and by-products to change. You will need to look carefully to see whether there are external factors for which you will need to normalize your data. Common ones to consider are total hours the process operated; total employee hours; area, weight, or volume of product produced; number of batches processed; area, weight, or volume of raw material purchased; and profit from product. For continuous processes, the product output or raw material input can be a good normalization factor. Flow processes may be measured by volume or weight, whereas plating or film-making may be better normalized by area.

In batch processes, production volume is usually related to waste production, but it may not be a linear relationship. For example, the quantity of solvent used at a printing plant is primarily a function of total volume of stock printed and ink used, but it is also significantly influenced by the number of color changes made.

Another difficulty in comparing production and waste quantities arises when the relationship is inverse. This situation occurs frequently when the production rate decreases to the point that age-dated input materials in the inventory expire. For some production processes, waste is generated during start-up and shut-down of equipment. The volume of waste created in both situations is inversely proportional to the production volume.

Revenue and profit factors can indicate the amount of activity but may not be reliable indicators if market forces often cause prices to change. Thus, monetary factors typically apply only to products in stable markets. Nonetheless, normalizing the waste stream by taking into account the facility's yearly production can, to a point, address market forces.

Step 8: Sustain the Pollution Prevention Program

The task of maintaining a viable pollution prevention program will be made easier with the establishment of a pollution prevention awareness program. Since such a program is often begun at the onset of the P2 program, this task involves mostly sustaining enthusiasm at a facility. Such a program is intended to promote employee involvement in the prevention effort. Since a significant amount of pollution prevention work has been performed up to this point, at least some facility employees should be familiar with the pollution prevention work that has been accomplished already. Because of this, an environmental manager should find it easier to maintain awareness of the program as a whole.

An environmental manager should continue to make all plant employees aware of the steps that have been taken, and any future plans. The objectives of the pollution prevention awareness program are to:

1. Raise awareness of environment-related activities at the facility.
2. Inform employees of specific environmental issues.
3. Train employees in their respective pollution prevention responsibilities.

4. Recognize employees for pollution prevention efforts.
5. Encourage employees to participate in pollution prevention.
6. Publicize success stories.

One of the most important elements of the waste minimization and pollution prevention awareness program is training. The training program should include all levels of personnel within the company. The goal is to make each employee aware of waste generation, its impact on the site and the environment, and ways in which waste can be reduced and pollution prevented.

A pollution prevention awareness orientation can be incorporated in the general orientation program given to all new employees. More detailed pollution prevention training should be provided to new employees after they have been on the job for a few weeks. This training will provide them with the skills they need to participate in pollution prevention. It also will emphasize the company's commitment to prevention.

At many plants, employees in certain jobs must be trained and tested on their knowledge of standard operating procedures specific to their work area prior to working there. Pollution prevention training can be incorporated into this standard training. It can also be incorporated into the quality assurance procedures qualification process.

Specialized training sessions on pollution prevention policy, procedures, and techniques should be provided to staff when their job scope is expanded or when they transfer to other areas within the company. These sessions should be considered part of the regular training program, and managers should have funds allocated to cover the costs.

If the progress of the pollution prevention program slows down, review the amount and type of pollution prevention training provided and consider increasing its frequency and audience. Periodic retraining of employees may be necessary when your policies and procedures change. Retraining employees will also reiterate the company's commitment to pollution prevention.

In addition to training, an environmental manager may want to consider an employee reward program. Progress in pollution prevention can be stated as an objective on which annual job performance evaluations are based, particularly at the management level. This delineates their responsibility for maintaining and enhancing the pollution prevention program. Using the formal mechanism of the written annual report to recognize efforts in this area raises the visibility of pollution prevention as something that is important to the company.

Employees who suggest pollution prevention measures that prove feasible and are slated for implementation should be publicized in the company newsletter or on bulletin boards. The estimated cost savings and other advantages to the company or unit should be included in this announcement. Periodic group meetings may be a good forum for announcing individuals' efforts to control pollution in the company's daily operations.

CASE STUDY

So how can a facility actually implement and sustain a successful P2 program? A facilitywide commitment to P2, careful process investigation and data gathering, and excellent internal communication are all important ingredients. A community newspaper in Minnesota managed to incorporate all these practices into a highly successful P2 program, reducing their waste output to the community landfill by over 25,000 pounds per year, and saving themselves over $18,000 per year. This case study gives excellent specific examples of the results of a functioning P2 program: effective brainstorming for creative waste-reducing ideas, careful measuring of waste outputs and costs, communication between different departments, and a collective commitment to reduce waste. Although the specific waste-reducing methods practiced by a newspaper staff may not apply to other industries, this case study illustrates the commitment that is required for any facility to reduce its waste.*

Source Reduction: A community newspaper case study

Facility: Herald Review
301 1st Avenue NW
Grand Rapids, Minnesota 55744

The Herald Review publishes a bi-weekly newspaper with a circulation of 8,000 and a weekly advertiser with a circulation of 20,000.

**Source*: Minnesota Office of Environmental Assistance, 1993.

This business made a commitment to source-reduce its waste as much as possible. Secondarily, what the staff members could not reduce they committed themselves to recycle.

As a result of their reduction measures alone, they are keeping 31 cubic yards and 25,150 pounds of waste out of their community's landfill each year. Not including the savings from avoided disposal fees, these measures result in a $12,914 yearly cost savings for the business.

Reduction is defined as any activity that reduces waste at its source. Employees examined their own wastestream and generated ideas to accomplish reduction. As they looked at their wastestream, they asked themselves the following questions.

- Where can I **reduce** the amount or the toxicity of material used to accomplish any task?
- Are there existing or new products I can **reuse** over and over again?
- Are there existing or new products that are repairable, refillable or more durable to give a **longer useful life?**

These are the three pillars on which they based their reduction program.

The publisher, Charlie Johnson, and the circulation manager, Ron Oleheiser, provided expertise and motivational staff leadership. Kenneth Brown, reduction specialist at the OEA, compiled this case study and facilitated staff meetings where recently implemented or potentially beneficial reduction measures were identified. The results of the businesses teamwork are contained within this report.

➲Use end rolls of paper stock

Newspaper stock arrives on large spools. When the paper on the spool is wound down, it is replaced so the spool does not run out of paper during a printing run. A local ceramic packaging firm buys the newspaper's left-over spools and uses the paper for packing material in place of styrofoam pellets, an excellent example of waste exchange.

Percentage reduction of this waste stream: 100%
None of this waste paper is thrown into the dumpster.

Volume of (compacted) waste avoided: 9.7 cu yds/year
This measurement is taken from a factory roll. This volume is extremely conservative because measurements were taken from new rolls, not disposal volume of used rolls.

Weight of waste avoided: 7,537 lbs/year

Cost savings before avoided disposal fees: $1,809 /year from safer
currently sold for 24¢ a lb.
$1,809 ÷ .24 = 7,537 lbs

Implementation of this measure resulted in a 6.4 percent savings in paper roll cost.

➲Reduce paper, ink, and labor waste caused by over-runs

The circulation manager made a concerted effort to determine the accurate count of publications needed for distribution. Overruns of 250 copies per publication are now decreased to no more than 50 overruns per publication. As a result, approximately 20 rolls of paper are saved each year.

Percent reduction of this waste stream: 80%

Volume of compacted waste avoided: 19 cu yds/year

20 rolls saved × [volume of cylinder = $\pi r^2 h$] 3.15 × 20"^{2} × 35" = 44,100 cu in ÷ 46,656 cu in for one cu yd = .945 cu yd/roll × 20 rolls = 18.9 cu yds. This volume is extremely conservative because measurements were taken from new rolls, not disposal volume of printed, folded newspaper.

Weight of waste avoided: 17,000 lbs/year

20 rolls saved × 850#/roll = 17,000

Cost savings before avoided disposal costs: $7,300 per year.

One roll of 35" wide paper costs $350 × 20 rolls saved each year = $7,000.
Estimate of ink cost savings: $45
Estimate of labor cost savings: $250
7,000 + 45 + 250 = 7,295

Implementation of this measure resulted in a 2.5 percent savings in publication paper cost.

➲Reuse of waste ink

Rather than dispose of excess colored and black ink, the staff collects it in catch pans and adds it to the respective hoppers. When a colored ink becomes too contaminated, it is added to the black ink hopper. Through this measure all ink is used.

Percent reduction of this waste stream: 100%

Volume of waste avoided: 250 gallons/yr

190 gallons/year of black ink
60 gallons/year of colored ink

Weight of waste avoided: 2,100 lbs/year

one gallon of ink equals approximately 8.5 lbs
1,615 lbs/year of black ink
510 lbs/year of colored ink
1,615 + 510 = 2,125 lbs

Cost savings not including avoided disposal fees: $2,615
$615.60 per year black ink
$2,000 per year colored ink
If this ink were not reused, it would be classified a hazardous waste.

Implementation of this measure resulted in a 17 percent savings in ink cost.

Issues: The newspaper staff expressed the desire to change from petroleum-based to nontoxic soy-based inks. Research showed that soy black ink is cost-competitive, but colored is not. Since combining inks is a source reduction requirement for the paper, when economically feasible the change will be made.

➲Reduce chemicals needed in film developing

The newspaper was using fresh chemicals for each photograph developing session. One gallon of developer was used each week. Since these chemicals were far from exhausted, the photography staff experimented by pouring this solution back into the bottle instead of down the drain. Through reuse, they found that life of developing solution was extended dramatically. One gallon of developer now lasts 3 weeks.

Percent reduction of this waste stream: 66%

Volume of waste avoided: 35 gallons/year
52 – 17 = 35

Weight of waste avoided: 210 lbs/year
6 lbs for each gallon × 35 gallons used per year = 210 lbs

Cost savings, not including avoided disposal fees: $140 /year
one gallon costs $4

$4 × 35 = $140

Implementation of this measure resulted in a 66 percent savings in developing-solution costs.

➲Roll own film

Since the newspaper works under a deadline, many unshot frames on pre-rolled film were often wasted when the film had to be developed. Now, the reporters roll their own film and load only the number of frames needed for the day. Through this measure, newspaper also switched from single use to refillable film canisters.

Percent reduction of this waste stream: 10%
The newspaper is using less film

Cost savings: $60 /year

Implementation of this measure resulted in a 10 percent savings in film costs.

Issues: Reporters take pleasure from the fact that they can roll the number of shots and film speed to suit the job at hand, without wasting material.

➲Reuse paste-up sheets

Ruled paste-up sheets are used to lay out newspaper pages. As a source reduction measure, the camera-ready copy and posters are peeled off the paste-up sheets so the paste-up sheets can be reused again and again. Sheets are reused an average of 6 times. Through this measure, use of over 6,700 sheets per year is cut to 1,000.

Percent reduction of this waste stream: 85%

Volume of compacted waste avoided: .4 cu yd/year
1,000 sheets = 6.5" × 25" × 20.5" = 3,331 cu in
5.7 × 3,331 = 18,986 cu in ÷ 46,656 cu in for one cu yd = .4 cu yd

Weight of waste avoided: 285 lbs/year
1,000 sheets = 50#
5.7 × 50# = 285

Cost savings not including avoided disposal costs: $570 /year
.10 cost per sheet × 5,700 sheets not purchased = 570
Labor cost to peel sheets is reported insignificant because staff uses the activity to fill down time that occurs during the production process.

Implementation of this measure resulted in a 85 percent savings in paste-up sheet costs.

➲ Reuse toner cartridges for computer printout and photocopier

Empty cartridges are refilled and rebuilt by a professional re-inker.

Percent reduction of this waste stream: 75%
Cartridges are refilled for reuse 3 times, then recycled.

Volume of compacted waste avoided: .49 cu yds/year
11" × 7" × 16.5" = 1,270 cu in × 18 = 22,869 + 46,656 = .49 cu yd

Weight of waste avoided: 27 lbs/year
1.5 lbs per unit × 18

Cost savings before avoided disposal costs: $900/year
Two replacement cartridges are required each month. New ones cost $105 each. 24 × $105 = $2,520
Rebuilding one costs $54.95 each.

Because new cartridges are now used once then reused 3 more times, new purchase is down to 6 cartridges per year. Reinking occurs 18 times.
6 [new] × $105 = $630 18 [refilled] × $54.95 = $989
$630 + $989 = $1,619–$2,520 = $901 saved

Implementation of this measure resulted in a 35 percent savings in cartridge costs.

➲Use narrow ruled reporter's notebooks

By switching from wide-ruled to narrow-ruled reporters' notebooks, the newspaper cut purchase of notebooks by half.

Percent reduction of this waste stream: 50%

Volume of waste avoided: .83 cu yd/year
Using 144 fewer notebooks each year

Weight of waste avoided: 34 lbs/year
2 lbs 13 oz per dozen × 12 = 33.75 lbs

Cost savings: $96 /year

Implementation of this measure resulted in a 50 percent savings in notebook costs.

➲Reuse of single-sided copy paper

The staff uses the remaining blank side of used paper for note pads and bundle labels.

Percent reduction of this waste stream: 100%
All single-sided paper is eventually used.

Volume of compacted waste avoided: .15 cu yd

Weight of waste avoided: 120 lbs

Cost savings not including avoided disposal fees: $180 /year.

As a result of implementing this measure NO paper is purchased for labeling, a 100 percent cost savings.

➲ Change to reuse cloth towels from disposable paper towels

There are four restrooms in the newspaper building. By changing to cabinets dispensing cloth rolled towels, the newspaper derived a financial as well an aesthetic benefit. Litter from the restroom floors disappeared, as did the large trash can. Because there was not room to hang the cabinet in the press room, one paper towel dispenser was kept there.

Percent reduction of this waste stream: 80%

Volume of waste avoided: .66 cu yds/year
Volume measurements were taken from new boxes of towels. Used towels take up significantly more volume. A new box of towels measures 18" × 12" × 12" = 3,888 cu in. × 8 (the number of boxes of towels kept out of the landfill) = 31,104 cu in ÷ 46,656 (for yds) = .66

Weight of waste avoided: 135 lbs/year

Cost savings not including avoided disposal fees: $120 /year

Implementation of this measure resulted in a 33 percent savings in towel costs.

➲ Use blank, excess mailing labels

The newspaper contracts for its mailing list to be put on mailing labels. There are often unused labels remaining on a partially used sheet. Newspaper staff decided to trim off these unused labels and use them on old manila file folders in need of new labels. The labels are used and the folders are given new life.

Cost savings is approximately $20 /year
From reduction in purchasing labels and does not including cost savings due to reused folders.

As a result of this measure NO labels are purchased, 100 percent cost savings.

➲ Coordinate reporters' activities

A concerted effort was made to plan reporters' itineraries so that stops were logical rather than haphazard.

Cost savings: $200 /year

Implementation of this measure resulted in a 16 percent savings in reporters' travel budgets.

➲ Reuse aluminum printing plates

5,000 24" × 36" aluminum sheets are used each year. The printing plates are sold for reuse as construction sheeting (waste exchange). Any not sold for reuse are recycled.

Volume of waste avoided: 1.1 cu yd of solid aluminum/year
1,000 plates are 12" × 24" × 36" = 10,368 cu in × 5 = 51,840 cu in/yr ÷ 46,656 cu" in one yd = 1.1 cu yds/yr

Weight of waste avoided: 4,250 lbs/year
1,000 sheets weigh 850 lbs, × 5 = 4,250 lbs/yr

Cost savings before avoided disposal fees: $1,200 per year in sales

Implementation of this measure resulted in a 14 percent savings in plate costs.

Issues: staff organized that current reuse and recycling of the plates was valuable, but due to the reduction program they pursued reuse a step further. The manufacturer was contacted to see if the plates could be taken back, re-buffed, and re-filmed so that newspapers could simply use them again. The outcome is unknown at this time, but it demonstrates the depth to which staff has integrated concepts of reduction. Because this measure has been in place some time, it was not included in the total cost, volume, and weight of waste avoided.

In addition

As a result of the newspaper's source reduction and recycling efforts, container services were decreased from three 12-cu-yd dumpsters per week (3 × 12 = 36 cu yd/wk) to three 30-gallon trash cans twice a week (3 × 30 × 2 = 180 gal × 231 cu" in one gal = 41580 cu"/wk ÷ 46656 cu"/yd = .89 cu yd/wk.) This is a 97 percent reduction in contracted hauling volume that cut garbage hauling expense by $3,900 per year.

Since much less time is now spent dumping office trash cans into the dumpster, cleaning service expense was consolidated, resulting in a further $134 savings each month or $1,608 per year. Total disposal and cleaning service savings is $5,508. When added to the $12,914 savings which result from source reduction measures, **the total waste management program saves over $18,400 /year.**

Note: The Herald Review has always used reusable mugs.

3

Industrial Sector Reports

This section describes common environmental problems for manufacturers in several Standard Industrial Classification (SIC) Codes. In addition, pollution prevention alternatives are given for many of the common problems. The nine SIC Codes presented in this section are illustrative of the manufacturing industry in general. Although the processes in different SIC Codes are often dissimilar at first glance, many pollution prevention problems are common to different industries. For this reason, useful pollution prevention techniques may be obtained from several of the following subsections, regardless of the particular industry under consideration.

To encourage pollution prevention, both general and company-specific descriptions of some pollution prevention advances are discussed. While the list is not exhaustive, it does provide information that can be used as the starting point for facilities interested in beginning their own pollution prevention projects. This section provides summary information from activities that may be, or are being, implemented by manufacturers. Facility-specific conditions must be carefully considered when pollution prevention options are evaluated. How each option affects air, land, and water pollutant releases must be examined in order to understand the full impacts of any potential change.

SIC CODE 20: FOOD AND KINDRED PRODUCTS

Environmental Impacts

Facilities in SIC Code 20 occupy a unique position relative to other types of industries' quantities of hazardous materials. The biggest concern for these facilities is Biological Oxygen Demand (BOD) and Chemical Oxygen Demand (COD) levels. High levels of BOD are caused by the use of oils and other organic material, which, usually through clean-up, tend to surface in wastewater streams. Since local Publicly Owned Treatment Works (POTWs) levy heavy surcharges to customers with high levels of BOD and COD, reducing those levels offer economic advantages. Some other areas of concern for SIC Code 20 facilities are ammonia emissions from refrigeration systems, "off-spec" products, lubrication oils from equipment (such as conveyors), cleaning materials, particulates (or fines), odors, waste containers (such as cans), corrugated paper, labels, and sanitizer and extraction solvents.

One component of the food industry, bakeries, produces significant quantities of hazardous and nonhazardous chemicals including carbon dioxide, ethanol, and ammonia. The fermentation process yields the end-products of carbon dioxide and ethanol. The rate of ethanol generation depends on dough temperature, quantity of sweetener, and type of yeast. No controls or process modifications are currently used to reduce ethanol emissions from bakeries.

BOD levels are typically high for bakeries. There are several factors that, when combined, lead to these excessive levels. The primary action responsible for the high BOD levels is poor housekeeping. There are several sources of organic matter discharged into the wastewater, including:

- Washdown of floors after shifts
- Rinsing of spent totes used in dough and sugar icing
- Washdown of fat- and oil-containing vessels and spent transfer lines.

The primary wastestream from the dairy industry is cheese whey. However, lactose, a major component of the whey, is valuable feedstock for fermentation processes. Approximately 45% of the whey produced in the United States is sent to municipal wastewater treatment facilities or disposed of in landfills.

Pollution Prevention Options

Pollution prevention often seeks nonhazardous substitutes for hazardous materials. Salt, which is both corrosive and potentially toxic to vegetation and aquatic life, can be replaced in road deicing with calcium magnesium acetate (CMA), which can be made from cheese whey. However, CMA currently costs $650/ton, which is 20 times more expensive than sodium chloride.

Industrywide

The following is a list of pollution prevention and waste minimization methods common to many facilities in SIC Code 20:

- Use an industrial-capacity vacuum system, whereby floors can be vacuumed of organic wastes before the final washdown.
- Oils, such as vegetable oil, should be segregated from the main wastewater and used for boiler fuel or collected as a separate waste.

- Use waste organics as animal feed.
- Recover as much product as possible.
- Reuse and recycle as much water as possible throughout the plant.
- Segregate waste streams.
- Develop a leak prevention program for valves, pumps, piping, and equipment.
- Drain or squeegee all raw material and product from tanks and vats before cleaning.
- Use dry cleanup methods. Avoid hosing material into drains if possible.
- Monitor the treatment plant for BOD on a weekly basis.
- Use water sparingly in hosing and cleaning operations (use High-Pressure-Low-Volume water sprays).
- Use valves that shut off automatically when water is not needed.

Dairy industry

In addition to the industrywide recommendations, the following is a list of pollution prevention and waste minimization methods that are common to the dairy industry:

- Segregate all major sources of waste with separated drains.
- Segregate or recycle nonpolluted water.
- Use the automation system of the plant to eliminate waste.
- Collect the first rinses from tanks and cleaning operations, and save the water-milk mixtures.
- Improve the efficiency of carbon-in-pulp (CIP) and sanitizing operations to reduce required concentration of these materials.
- Use postcleaning rinses as make-up water for sanitizing and cleaning.

- Collect lubricants and reclaim for use.
- Mark all valves clearly, especially multiport, so that it is practically impossible for inexperienced help to turn the valves the wrong way.
- Handle with extreme care all sanitary fittings, valves, rotary seals, pump parts, and filler parts during every phase of operation to prevent marring, which may cause leaks.
- Provide accurate temperature controls on plate, tubular, or surface coolers to prevent freezing-on, especially with glycol systems.
- Install suitable liquid level controls with automatic pump stops or other devices at all points where overflows are likely to occur.
- Never fill cheese vats, ice cream mix vats, or cooling tanks to such a high level that spillage will occur when the product is agitated.
- Worn-out or obsolete equipment or parts of equipment, including sanitary valves, fittings, and pumps, should be repaired or replaced.
- Buckets should be readily available so that should a leak occur that cannot be repaired during processing, the product is not allowed to go down the floor drain.
- When unloading tank trucks, avoid spillage from the sanitary pipe or hose connections. Be certain the tank is empty before disconnecting the sanitary pipe or hose.
- Do not use a constantly running water hose in any room.
- Eliminate the cause of spillage, rather than just wash it away after it has occurred.
- Where sugar is used in the manufacture of dairy products, take care that it is not spilled on the floor and washed into the sewer. Sugar will pollute the stream just as much as milk solid.

- Provide standby pumps for pumping whey from cheese vats to storage tanks in case of power or pump failure.
- Sweep up all spilled cheese curd particles from the floor. Do not wash them down the sewers.
- Pipelines should be installed so that they are properly supported to eliminate vibration-induced leaking joints (where welded lines are not used or points in welded lines where CPI gaskets are used) and to ensure that the lines are properly pitched to ensure draining of the lines.
- For plastic and glass bottle fillers, cappers should be maintained in first-class condition to avoid breakage and product loss.
- Paper-filling machines should be maintained in order to avoid jams and product spillage.
- Fillers should be equipped with drip shields and the spills collected to avoid product going to drain.
- Filler valves should be checked to see that all containers are filled to correct capacity at the temperature filling.
- Bottles and paper containers should be handled carefully during casing, stacking, loading, and delivery to avoid product losses to drain.
- Adequate weirs and continuous sampling equipment should be provided at the outlet of all dairy plants for monitoring wastes concentrations and volumes.
- Drain all lines, tanks, and processing vats before rinsing.
- Rinse before product dries on surface.
- Reduce losses from high-volume pasteurizing units through recycling.
- Reduce losses through continuous processing.
- Fit bag type filters in the floor drains in the cottage cheese room.

SIC Code 22: Textiles

Environmental Impacts

Traditionally, the textile industry is very energy-, water-, and chemical-intensive. Approximately 60% of the total energy consumed is used within the wet-processing stages. Most wet processing involves treatment with chemical baths, which often require washing, rinsing, and drying steps between key treatment steps. Consequently, large volumes of wastewater are generated with a diverse range of contaminants, which must be treated prior to disposal. Much energy is consumed to heat and cool chemical baths and washwater and to dry fabrics or yarns. Additionally, there are a few problems of air pollution by chemicals and lint, but these are minor.

When the complete textile product life cycle is considered, it is apparent that chemical suppliers (for fibers, dyes, and auxiliaries) have a tremendous influence on the environmental impact of textile manufacturing. In fact, there is increasing pressure from the textile mills on their suppliers to improve product performance via increased exhaustion and efficiency of dyes, less-toxic and fewer auxiliary chemicals needed, and less-volatile and permanent finishes.

Waste Effluents

Wastes in textile mill effluents can be divided into the following categories:

- Naturally occurring dirts, salts, oils and greases on cotton and wool
- Chemicals added or removed during the many different process operations
- Fibers removed by chemical and mechanical action during processing

These wastes typically leave the mill in the wastewater. Types of materials that may be expected to predominate in textile wastewater include:

- Metals
- Nondegradable surfactants
- Toxic organic materials, such as phenols, aromatic solvents, methyl chloride, perchloroethylene (PERC), and oxalic acid

Generally, wastewater is colored, highly alkaline, high in BOD and COD, and at elevated temperature. Of particular concern are compounds that are toxic to aquatic life, such as heavy metals (primarily from dyeing and finishing) surfactants, and compounds used throughout the wet-processing steps.

The BOD load comes from several sources. Synthetic oils and waxes include winding emulsions, paraffin wax, knitting oil, coning oil, and other lubricants. These materials have significant BOD themselves, and the emulsifiers used to remove and suspend them in preparation also contribute. Dyeing processes vary greatly in the amount and type of wastes produced, but the primary sources of BOD from continuous dyeing are the soap-off chemicals (surfactants) and the dyeing auxiliaries used in the pad bath, which may wash off. Typically scouring wastes contribute a large portion of the total wastes but less than 50% of BOD loads in wastestreams from the preparation processes. Problems with spent dye liquors from batch dyeing are difficult to handle because wastestreams are usually dumped into common pits or trenches.

Air Emissions

Air emissions emanate from several places in typical textile operations:

- Hot air dryers
- Dyeing machines
- Storage tanks
- Warehouse area
- General ventilation (fugitive emissions)

Some of the solvents used in textile manufacturing that are of concern include:

- Perchloroethylene (or tetrachloroethylene)
- Trichloroethylene (TCE)
- Methylene chloride (or dichloromethane)
- 1,1,1-Trichloroethane
- Toluene (or Toluol)
- Benzene (or Benzol)
- Xylene (or Xylol)
- Ethylene dichloride
- Varsol (or Naptha)

These solvents are typically used in the dry cleaning of 100% synthetic fiber knit fabrics, and woven and wool fabrics; in specialty operations such as tricot and solvent scouring; in dyeing operations; and in some finishing operations for impregnation or coating of textile fibers. In addition, solvents are used to clean machinery such as rollers and spinning machines used in textile manufacturing.

Other Hazardous Chemicals

Maintenance chemicals are sometimes overlooked as a source of hazardous chemicals. Although the typical textile mill usually has an approval procedure and quality control for raw materials used in processing, sometimes maintenance chemicals are not included in the approval procedure because they do not affect the mill's direct cost and quality control. They are also purchased and used less frequently than process chemicals. Typical contents include methylene chloride, perchloroethylene, and other hazardous materials.

Other types of nonprocess chemicals that have been known to contribute to toxicity of wastewaters include biocides and herbicides. Biocides are frequently used in cooling towers and nonprocess water purification and processing. Biocides are also used in certain applications such as finishes for hosiery, tents, and awnings. Herbicides are used to control grass, weeds, and other vegetation near bulk storage tanks.

Pollution Prevention Options

The primary focus for advancing pollution prevention in textiles manufacturing is on reducing water, chemicals, and energy use and on using less-hazardous, more-efficient and more-recoverable compounds. These goals are accomplished through material substitution, process changes, and reuse strategies.

Source reduction typically results in important economic benefits for manufacturers in several ways including decreased liability for waste, decreased loss of materials and energy waste, decreased waste collection costs, decreased handling and treatment costs, and potential recycle and recovery value of waste.

Some ideas that have been used commercially for pollution source management are as follows:

Preparation	*General*
Recovery systems	Waste characterization
Waste stream reuse	Raw materials
Alternative processes	Equalization

Finishing	*Dyeing*
Substitution	Chemical substitutions
Alternative processes	Alternative processes
Reuse	Reconstitution/reuse of dyebaths

Source reduction may be accomplished by several different methods and techniques:

- Eliminating phosphates wherever possible
- Saving energy by removing water by centrifugal means, rather than evaporation
- Evaluating all maintenance chemicals for metals and toxicity
- Conservation of chemicals
- Process modifications
- Improved maintenance
- Waste recovery
- Separate wastestreams
- Using liquid-ammonia mercerization instead of caustic
- Using bathless washers
- Replacing all chromates used for oxidation of vat dyes
- Optimization of chemicals
- Equipment modifications
- Improved housekeeping
- Recycling and reusing waste
- Chemical conservation

See also Table 3.1.

Waste Reuse and Exchange

Several methods have been used to recover and segregate wastewater streams. Once-through, noncontact

Table 3.1. Material Change Strategies in Wet Processing

Option	Example	Benefit
Alternative desizing agents	H_2O_2 rather than enzyme for desizing of starch	Resulting products of CO_2 and H_2O over-hydrolyzed starch-using enzymes
	Enzymes that degrade starch to ethanol	Reduced BOD of spent desizing bath and opportunity to recover ethanol as fuel
Bleaching agent Alternate dyes	Copper-free for producing green shades in 100% cotton fabric	Reduced toxicity (metal content) of spent dye bath and washwater
	Improved fixation reactives such as with Remazol and others	Less unreacted and hydrolyzed (degraded) dye in spent bath and washwater improves reuse opportunities
	High-temperature reactive	Energy reduction, elimination of caustic bath required after disperse dyeing
Substitute auxiliaries (all processes)	Substitute for phosphates such as acetic acid (pH control)	Reduction in phosphorus load in wastewater
	New washing agents (Sandpure RSK and others)	Increased wash efficiency, decreased water consumption

continued

Table 3.1. (continued)

Option	Example	Benefit
Pad-patch dyeing	Transfer of dye to cotton, rayon, and blended goods through rollers (continuous method)	Reduction in water (2 gal/lb vs. 20 on becks), energy (2000 vs. 9000 BTU/lb) and chemicals use, increased productivity
Foam technology	Application of dye through foam media (air dispersed in liquid) or other solvents for dyeing and printing	Reduced water and energy consumption, chemical waste, and time necessary for drying
Spraying technology	Application of finishes using sprays	Reduced energy and water usage, and chemical waste (most finish remains on fabric)
Washing technology	Countercurrent washing, vibrating-reed jet washers, mechanical means for increased turbulence	Improved washing efficiency, thus decreased water and energy usage
Process consolidation	Single-pad-steam-wash sequence using unique combinations of chemicals	Decreased water and energy usage, salt in effluent, and process times

cooling water can be reused by sending it back to a clear well or influent water line to the mill, which can result in significant water savings.

There will always be circumstances in which wastes are unavoidable. In these cases, wastes can become marketable. For example, waste print paste has significant potential for improved market value as a concrete additive for colored concrete and lint can be sold as a recycled feed stock.

Waste exchanges turn one industry's waste products into raw material in another manufacturing operation. There are essentially two types of waste exchanges: passive (information) and active (material). In a passive waste exchange, generators list in a catalog the wastes they have available and potential users list the wastes they want. Information exchanges may also provide up-to-date waste recycling news, and make an active effort to find possible users for available wastes.

In an active exchange, a broker serves as a liaison between the generators and potential users of the wastes. The materials exchange may analyze and process the waste to make it a suitable raw material for buyers to use. An ongoing system of waste exchange among several companies that are in close proximity to one another is called "industrial ecology."

SIC Code 26: Paper and Allied Products

The mills in the paper and pulp industry, SIC Code 26, are classified according to their primary product: pulp, paper, and paperboard. Pulp mills produce pulp, primarily from wood, but also from wastepaper, rags, linters, and straw. Paperboard mills take pulp or secondary fibers (recycled from previous paper products) and manufacture paperboard.

Secondary fibers are used throughout the industry. The quality of the fibers establishes their use. High-quality fibers are used primarily for newsprint, whereas low-quality fibers may be used in corrugated papers. The final product determines the level of de-inking, washing, and bleaching that is needed.

Environmental Impacts

From the wood yard through pulping and paper making, all of the processes have environmental impacts and opportunities for pollution prevention. For example, debarking logs by the wet process uses 2,000 to 3,000 gallons of water per ton of wood and contributes biological oxygen demand (BOD) and total suspended solids (TSS) to the wastewater. Pulp bleaching creates furans and chlorinated compounds, such as dioxin, and produces significant quantities of hazardous waste. Large volumes of wastewater are generated as a result of the pulping, bleaching, and associated processes. The water contains dissolved lignins, suspended solids, adsorbable organic halides (AOX), and several other organic compounds. Several well-known recycling techniques have dramatically decreased the need for freshwater and loads on waste treatment facilities.

Pollution Prevention Options

Wood Yard Operations

The wood yard turns logs into chips. The following alterations can improve plant efficiency and lower emissions. Switching from wet to dry debarking would reduce water consumption and wastewater production. Furthermore, the wastes (bark) would not have to be dewatered. They could be fed directly into the energy recovery boilers.

The efficiency of the operation can be improved by controlling the size of the chips. Properly maintaining the chipper machinery increases chip uniformity. The chips can be screened for thickness and size, and the oversized material can be shunted back to the chipper. Without screening, oversized chips enter the digester, but the kraft liquor does not penetrate them completely. Chips that fail to digest are typically fed to hogged-fuel boilers, which contribute 60% to 80% of total reduced sulfur (TRS) emissions. Chipping and screening the wood to the correct size would eliminate this source of pollution and would also conserve bleach and reduce its associated emissions.

To prevent chips and dust from entering the wastewater treatment plant via stormwater runoff, dikes should be built around the chip piles and processing areas. By avoiding surface water contamination, this simple method may eliminate the need for a separate stormwater permit.

Pulping and Chemical Recovery

Lignin is the "glue" that sticks the wood fibers together, and pulping is the process of ungluing and releasing the fibers. Many pollution prevention methods are available to improve the pulping process and the recovery of the chemicals that are used.

Several methods relate directly to lignin removal. For instance, pretreatment of the pulp with nitrogen dioxide (NO_2) allows up to 80% lignin removal in the oxygen stage, which is normally limited to 50%. Demethylation (using potassium thiophenoxide in diethylene glycol at 200°C) also enhances the delignification process by providing more active sites on the pulp. Anthraquinone can be added to the digester to accelerate the pulping process and increase yield. Using polysulfides has been shown to be superior to typical kraft cooking. Simply extending the time of the delignification process results in greater lignin removal and reduces the demand on the bleaching process.

Other pollution prevention methods improve the recovery of the chemicals used in the pulping process. Closed screen rooms, for example, increase the efficiency of the chemical recovery, which lowers the demand for virgin chemicals to make up the white liquor. Improving the brownstock washing will increase chemical recovery and reduce VOC and odorous emissions. Installing a lime mud diversion basin will improve resource recovery while reducing sludge accumulation, wastewater loading, and treatment cost.

Some of the chemicals recovered can be used in other parts of the plant. Washwater from the causticizing system can be recycled as a scrubber solution for the smelt-dissolving tank, bleach plant and chlorine dioxide plant vents, lime slakers, and lime kilns. Emissions from the high-volume-low-concentration and low-volume-low-concentration vents can be burned for their heat content. Steam stripping of foul condensates, which can reduce VOC emissions and lower BOD loads downstream, produces stripper overheads that can then be burned for energy recovery and bottoms that can be recycled in brownstock washing.

The remaining pollution control methods target emissions directly. Maintaining equipment (e.g. valves, pumps, and tanks) can avoid spills and leaks and reduce fugitive emissions. The formation of chlorinated compounds and contaminant loading in wastewater sludges can be lessened by using appropriate defoamers and pitch dispersants. Conversion of recovery boilers to low-odor designs can reduce emissions by more than 90% while improving boiler efficiency.

Pulp Bleaching

Pulp is bleached to convert it from brown to white. The bleaching process contributes significant amounts of hazardous waste to the environment. The lignin remaining on the fibers after the pulping process colors the fibers brown. This residual lignin is removed during bleaching. The introduction of oxygen—alone or together with peroxides and hypochlorites—can enhance lignin extraction. Not only is this method relatively inexpensive, but it also reduces BOD, COD, and color in the effluent.

Most of the pollution prevention methods for pulp bleaching target the formation of chlorinated compounds by substituting reagents or using chlorine more efficiently. Using ozone in place of chlorine as a bleaching agent, followed by oxygen delignification, eliminates the formation of chlorinated compounds. The substitution results in lower wastewater loadings of BOD, color, and adsorbable organic halides (AOX). The formation of chlorinated compounds in the first bleaching stage can be minimized by substituting chlorine dioxide for elemental chlorine while carefully controlling the Kappa factor, keeping it below 0.15 with special instrumentation that regulates chemical inputs. Chlorine dioxide can be substituted for hypochlorite. High-grade chlorine dioxide is the most effective and reduces the formation of chloroform.

To conserve on the use of chlorine and minimize overchlorination, the chlorine should be added at intervals in several batches, rather than all at once. Localized overchlorination can be avoided by using high-shear equipment, which promotes uniform solutions and the efficient use of the bleach.

The final method for the bleaching process is not targeted specifically to chlorine use. Countercurrent, jump-stage washing diminishes water demands and treatment facility loads and saves energy.

Pulp Drying and Paper Making

The pulp-drying and paper-making processes offer opportunities to increase yield, reduce water use and subsequent treatment, save energy, and substitute less hazardous chemicals. Several techniques can recover fibers from whitewater and brownwater, thereby increasing yield and reducing BOD. Techniques using drums, settling tanks, flotation, or polydisk savealls can recover more than 90% of the fibers. The water can be recycled back into the main or auxiliary processes.

Using cascade systems or recirculating water systems to operate water ring vacuum pumps can lessen water and energy demands further. Water can be recovered by collecting the steam condensates from pulp dryers and paper machines. Pulp drying and paper making processes use a number of chemicals: flocculents, sizing agents, fillers, resins, brighteners, pigments, and dyes, which should be examined to determine if less hazardous substances can be substituted.

SIC Code 27: Printing and Publishing

The printing and publishing industries include facilities employed directly in printing, those firms that perform services for the printing trade (e.g., typesetting, electrotyping, engraving, photoengraving, and bookbinding), and companies engaged in publishing books, newspapers, and periodicals, regardless of whether or not the publishers do their own printing.

Printing Processes and Their Uses

Ranked according to their market share, the most common printing processes are lithography, gravure, flexography, letterpress, and screen. The lithographic process is used to print books, posters, greeting cards, labels, packaging, advertising flyers, brochures, and periodicals, and to reproduce artwork. Gravure is used for printing high-quality publications, magazines, catalogs, and advertising. It is best suited for large volume runs. It is also used to print polyolefin packaging, paperboard boxes, and labels. Flexography is employed to print packaging, such as plastic wrappers, corrugated boxes, milk cartons, foil, and paper bags, and also for shower curtains. Web letterpress is used for newspapers and magazines. Screen printing is used for specialty printing, T-shirts, posters, banners, decals, and wallpapers. It is also used to print patterns on electronic circuit boards prior to etching.*

Environmental Impacts

All of the printing processes use raw materials (including natural resources and potentially hazardous substances), require energy, and generate wastes that may

---

*Robert Noyes, "Pollution Prevention Technology Handbook." Park Ridge, NJ: Noyes Publications, 1993.

adversely affect the air, water, and land. Certain chemicals involved in printing volatilize, which may contribute to smog formation; other chemicals may go down drains and impact freshwater or marine ecosystems; and solid wastes necessitate proper disposal. The cumulative impact of all printing operations on the environment is substantial, though the average small print shop, by itself, is not a major source of pollution.

Raw Materials

The principal raw materials used by the printing and publishing industry are substrates (e.g., paper, plastic, cloth, wood, and metal) and inks. Other inputs include photographic films, photoprocessing chemicals (developers, fixers, wash baths, reducers, intensifiers), printing plates, plate processing chemicals, gravure cylinders, lubricating oils, fountain solutions, cleaning solvents, and rags.

The enormous volume of paper consumed by printing and publishing links these activities to the paper industry (SIC 26), discussed earlier, and to environmental issues surrounding commercial forestry and the paper recycling industry. Demands placed by printers and publishers on the pulp and paper industry (e.g., brightness) determine to some extent what pollution prevention options the latter industry sees as viable. This is an example of a possible synergy between two industries that could work together to reduce their total pollution load.

Wastes and Emissions

The wastes reflect the inputs. By far the largest contribution to solid waste (at 98% for the lithographic industry) is spoiled paper and paper wrap. Most of this paper is recycled, incinerated, or disposed of as trash. The image processing, platemaking, proofmaking, printing, and finishing processes generate additional solid wastes, including film, empty containers, used blankets, and damaged products. Equipment cleaning adds to the wastestream items such as soiled rags.

Wastewater arises from image processing, platemaking, and printing processes. Air emissions result from several of the processes and from equipment cleaning. For example, most of the isopropyl alcohol (IPA) used in lithographic printing evaporates, contributing to ground-level ozone (smog) production. Isopropyl alcohol decreases the surface tension of lithographic fountain solutions, allowing the solution to adhere more readily to the nonimage area of the plate cylinder. However, newer, low-volatility fountain solutions use surfactants, glycols, and glycol ethers in place of the IPA to cut surface tension. Many printers are converting to non- and low-IPA fountain solutions.

The actual wastes generated by a printer will depend on the technologies and material used and whether waste reduction measures are employed. The following list indicates wastes and emissions that are commonly found in the printing and publishing industry and offers a starting point when looking for pollution prevention opportunities.

Hazardous Wastes

- Photographic wastes (intensifiers, scrap litho film)
- Spent solvents (1,1,1-trichloroethane, methylene chloride)
- Waste inks with solvents or heavy metals (inks and empty containers)
- Cleaning rags soaked with solvent or containing residues of heavy metals
- Strong acid or alkaline wastes (ammonium hydroxide, hydrochloric acid)

- Spent plating wastes (spent plating solutions and sludges)
- Ink sludge containing heavy metals (ink sludges with chromium, lead, or cadmium)
- Containers with hazardous residues of solvents, inks, or adhesives
- Miscellaneous (used oil from compressors, refrigeration tank filters)

VOCs

- Inks (xylene, ketones, alcohols)
- Fountain solutions (isopropyl alcohol)
- Adhesives (binding, labels)
- Cleanup (toluene, naphtha, kerosene, methanol)

Hazardous Wastewater Emissions

- Rinse from photo developing processes
- Any liquid hazardous waste dumped down drain

Nonhazardous Solid Wastes

- Waste substrate (paper, plastic, foil, rejects)
- Nonhazardous waste inks (inks and empty containers)
- Rags and wipes without solvents or hazardous residues
- Empty cartons, wrappers, and roll cores

Pollution Prevention Options

A systematic look at each operation will show where the wastes are generated and where the greatest potential for pollution prevention lies. For each of the operations of the printing industry, several pollution prevention methods are listed below.

Material Handling and Storage

- Proper storage of materials
- Material preinspection
- Inventory control
- Restrict traffic through area
- Recycle empty containers
- Purchase quantities according to needs
- Recycle photographic film and paper

Image Processing

- Material substitution
- Electronic imaging and laser platemaking
- Extend bath life
- Recover silver and recycle chemicals
- Employ countercurrent washing

Plate Processing

- Reduce solution loss
- Use nonhazardous developers and finishers
- Replace metal etching/plating operations

Make-ready

- Ink/water ratio sensor
- Computerized registration
- Automatic ink key setting system

Printing

- Collect and reuse solvent
- Store ink properly
- Alternative printing techniques
- Use less hazardous inks
- Use alternative cleaning solvents

Finishing

- Reduce paper use and recycle waste paper

SIC Code 2911: Petroleum Refining

Petroleum refineries play an important role in the U.S. economy, supplying approximately 40% of the total energy used in the United States and virtually all of the energy consumed in the transportation sector. The petroleum refining industry, SIC 2911, is one of several subclassifications of the petroleum industry, SIC 29. The petroleum refining industry includes the production of petroleum products through distillation and fractionation of crude oil, redistillation of unfinished petroleum derivatives, cracking, or other processes. The related industries under SIC 29 are 2951, Asphalt Paving Mixtures and Blocks; 2952, Asphalt Felts and Coatings; 2992, Lubricating Oils and Greases; and 2999, Petroleum and Coal Products, Not Elsewhere Classified. Certain products that are produced by the petroleum refining industry are also manufactured by other industries, including 2865, Cyclic Organic Crudes and Intermediates, and Organic Dyes and Pigments; 2869, Industrial Organic Chemicals; 2819, Industrial Inorganic Chemicals, Not Elsewhere Classified; and 2821, Plastic Materials.

Environmental Impacts

Raw material input to petroleum refineries is primarily crude oil; however, petroleum refineries use and generate an enormous number of chemicals, many of which leave the facilities as air emissions, solid waste, or in wastewater. Pollutants generated typically include VOCs, carbon monoxide (CO), sulfur oxides (SOx), nitrogen oxides (NOx), particulates, ammonia (NH_3), hydrogen sulfide (H_2S), metals, spent acids, and numerous toxic organic compounds. When discussing material outputs of the petroleum refining industry, it is important to distinguish between the outputs of the industry itself and the outputs resulting from the use of refinery products. The pollutant outputs from the refining facilities are modest in comparison to the pollutant outputs realized from the consumption of petroleum products by the transportation sector, electric utilities, chemical manufacturers, and other industrial and commercial users.

Wastewater

Wastewaters consist of cooling, process, sanitary sewage, and storm waters. Wastewaters are treated in on-site wastewater treatment facilities and then discharged to POTWs or discharged to surface waters under NPDES permits. In addition, some facilities use underground injection of some wastewater streams. Many refineries unintentionally release, or have unintentionally released in the past, liquid hydrocarbons to ground water and surface waters. At some refineries contaminated ground water has migrated off-site and resulted in continuous "seeps" to surface waters. While the actual volume of hydrocarbons released in such a manner is relatively small, there is the potential to contaminate large volumes of ground water and surface water, possibly posing a substantial risk to human health and the environment.

Pollution Prevention Options

Establish a leak detection and repair program. Fugitive emissions are one of the largest sources of refinery hydrocarbon emissions. A leak detection and repair (LDAR) program consists of using a portable VOC detecting instrument to detect leaks during regularly scheduled inspections of valves, flanges, and pump seals. Leaks are then repaired immediately or are scheduled for repair as quickly as possible. A LDAR program could reduce fugitive emissions 40% to 64%, depending on the frequency of inspections.

Place secondary seals on storage tanks. One of the largest sources of fugitive emissions from refineries is storage tanks containing gasoline and other volatile products. These losses can be significantly reduced by installing secondary seals on storage tanks.

Reduce the generation of tank bottoms. Tank bottoms from crude oil storage tanks constitute a large percentage of refinery solid waste and pose a particularly difficult disposal problem due to the presence of heavy metals. Tank bottoms are composed of heavy hydrocarbons, solids, water, rust, and scale. Minimization of tank bottoms is carried out most cost effectively through careful separation of the oil and water remaining in the tank bottom. Filters and centrifuges can also be used to recover the oil for recycling.

Control solids entering sewers. Solids released to the wastewater sewer system can account for a large portion of a refinery's oily sludges. Solids entering the sewer system (primarily soil particles) become coated with oil and are deposited as oily sludges in the API oil/water separator. Because a typical sludge has a solids content of 5% to 30% by weight, preventing one pound of solids from entering the sewer system can eliminate 3 to 20 lbs. of oily sludge.

Minimize solids leaving the desalter. Solids entering the crude distillation unit are likely to eventually attract more oil and produce additional emulsions and sludges. The amount of solids removed from the desalting unit should, therefore, be maximized. A number of techniques can be used, such as using low shear mixing devices to mix desalter washwater and crude oil; using lower pressure water in the desalter to avoid turbulence; and replacing the water jets that are used in some refineries with mud rakes, which add less turbulence when removing settled solids.

Control heat exchanger cleaning solids. In many refineries, using high pressure water to clean heat exchanger bundles generates and releases water and entrained solids to the refinery wastewater treatment system. Exchanger solids may then attract oil as they move through the sewer system and may also produce finer solids and stabilized emulsions that are more difficult to remove. Solids can be removed at the heat exchanger cleaning pad by installing concrete overflow weirs around the surface drains or by covering drains with a screen. Solids generation can also be reduced by using antifoulants on the heat exchanger bundles to prevent scaling.

Control of surfactants in wastewater. Surfactants entering the refinery wastewater streams will increase the amount of emulsions and sludges generated. Surfactants can enter the system from a number of sources, including washing unit pads with detergents; treating gasolines with an end point over 400° F, thereby producing spent caustics; cleaning tank truck interiors; and using soaps and cleaners for miscellaneous tasks. In addition, the overuse and mixing of the organic polymers used to separate oil, water, and solids in the wastewater treatment plant can actually stabilize emulsions. The use of surfactants should be minimized by educating operators, routing surfactant sources to a point downstream of the DAF unit and by using dry cleaning, high-pressure water, or steam to clean oil surfaces of oil and dirt.

Eliminate use of open ponds. Open ponds used to cool, settle out solids, and store process water can be a significant source of VOC emissions. In many cases open ponds can be replaced with closed storage tanks. In cases where the VOC content represents product, vapors from the tank can be captured and returned to the product mix.

Replace old boilers. Older refinery boilers can be a significant source of SOx, NOx, and particulate emissions. It is possible to replace a large number of old boilers with a single new cogeneration plant with emissions controls.

Reduce the use of 55-gal. drums. Replacing 55-gal. drums with bulk storage can minimize the chances of leaks and spills.

Install rupture discs and plugs. Rupture discs on pressure-relieve valves and plugs in open-ended valves can reduce fugitive emissions.

Install high-pressure power washer. Chlorinated-solvent vapor degreasers can be replaced with high-pressure power washers, which do not generate spent solvent hazardous wastes.

Identify benzene sources and treat benzene before it enters the wastestream.

Recycling Options

Recycle and regenerate spent caustics. Caustics used to remove hydrogen sulfide and phenol contaminants from intermediate and final product streams often can be recycled. Spent caustics may be saleable to chemical recovery companies if concentrations of phenol or hydrogen sulfide are high enough. Process changes in the refinery may be needed to raise the concentration of phenols in the caustic to make recovery of the contaminants economical.

Recycle phenols on-site. Phenols can also be recycled on-site by reducing the pH of the caustic until the phenols become insoluble, thereby allowing physical separation. The caustic can then be treated in the refinery wastewater system.

Use oily sludges as feedstock. Many oily sludges can be sent to a coking unit or the crude distillation unit, where they become part of the refinery products. Sludge sent to the coker can be injected into the coke drum with the quench water, injected directly into the delayed coker, or injected into the coker blowdown contactor used in separating the quenching products. Use of sludge as a feedstock has increased significantly in recent years and is currently carried out by most refineries. The quantity of sludge that can be sent to the coker is restricted by coke-quality specifications, which may limit the amount of sludge solids in the coke. Coking operations can be upgraded, however, to increase the amount of sludge that they can handle.

Material Substitution

Use nonhazardous degreasers. Spent conventional degreaser solvents can be reduced or eliminated through substitution with less-toxic or biodegradable products.

Eliminate chromates as an anticorrosive. Chromate-containing wastes can be reduced or eliminated in cooling tower and heat exchanger sludges by replacing chromates with less-toxic alternatives, such as phosphates.

Use high-quality catalysts. By using catalysts of a higher quality, process efficiencies can be increased while the required frequency of catalyst replacement can be reduced.

Replace ceramic catalyst support with activated alumina supports. Activated alumina supports can be recycled with spent alumina catalyst.

SIC Code 30: Rubber and Miscellaneous Plastics

Environmental Impacts

Four general types of pollution and resource outputs can occur at one or more stages of the plastics manufacturing process. Disposal presents an additional concern. Manufacturing outputs include spills, leaks, and fugitive emissions of chemicals during the application of additives prior to molding or during finishing; wastewater discharges during cooling and heating, cleaning, and finishing operations; plastic pellet releases to the environment prior to molding; and fugitive emissions from molding and extruding machines.

Chemicals

The chemicals added prior to molding and during the finishing of plastic products may be released by spills during weighing, mixing, and general handling through leaks from chemical containers and molding machines, or as fugitive dust emissions from open chemical containers. The chemicals used in plastic product manufacturing are usually added in such small amounts that most manufacturers do not consider them to be a problem; however, some of the additives could be toxic, and therefore even small amounts could present significant problems.

Wastewater

Contaminated wastewater is another concern in the plastic product industry. Water used in the plastic molding and forming processes falls into three main categories: (1) water to cool or heat the plastic products; (2) water to clean the surface of both the plastic products and the equipment used in production; and (3) water to finish the plastics products. The only pollutant found in a treatable concentration in some wastewater discharged after contact cooling and heating is bis(2-ethylhexyl) phthalate (BEHP), and it is toxic.

Pellet Release

The release of plastic pellets into the environment is of concern because these pellets may end up in wastewater. Plastic pellets and granules used to mold intermediate and final plastics products are often swept from the floor during transport or while being loaded into molding machines. Although they are inert, plastic pellets can cause harm if runoff carries them to wetlands, estuaries, or oceans, where they may be ingested by seabirds and other marine species. EPA stormwater regulations classify plastic pellets as "significant materials," and, therefore, a single pellet in stormwater runoff is subject to federal regulatory action.

Fugitive Emissions

Fugitive emissions from the molding and extruding processes may be an environmental concern because of the many additives, including cadmium and lead, which can be released during the application of high heat and pressure. Trade association officials (i.e., American Plastic Council and the Society of the Plastic Industry) are currently researching the composition of these emissions and their possible effects on worker safety and air quality.

Environmental Impacts of Rubber Manufacturing

The compounding area, where dry chemicals are weighed and put into containers before they are mixed, can be a source of fugitive emissions, spills and leaks. Because additives must be preweighed, some facilities allow the chemicals to sit in large open bins on scales, a situation that increases the potential for fugitive dust emissions. Most mixing facilities have eliminated this problem by purchasing their chemicals in small, preweighed, sealed polyethylene bags. The sealed bag is placed directly into the banbury mixer. If chemicals are not in preweighed bags, fugitive emissions are also produced as the chemicals are loaded into the mixer. Emissions from the internal mixers are typically controlled by baghouses. Exhausts from the collection hoods are ducted to the baghouses for control of particulate and possibly particle-bound semi-volatiles and metals.

VOC Emissions

VOC emissions are an environmental concern in rubber manufacturing. Solvents are used to degrease equipment and tools and as a type of adhesive during building. Typically, releases of solvents occur either when the spent solvent solutions are disposed as hazardous wastes or when degreasing solvents are allowed to volatilize. In some facilities, mold release compounds, sprayed onto the cavities of compression molds, produce significant fugitive emissions.

Solid Waste

Solid wastes are also an issue for rubber product manufacturers. Surface grinding activities that generate dust and rubber particles are typically controlled by a primary cyclone and a secondary baghouse or electrostatic precipitator. This baghouse-captured particulate matter (e.g., chemicals and ground rubber) from compounding areas, banburys, and grinders is a source of solid waste.

Scorched rubber from mixing, milling, calendering, and extruding is a major solid waste source within rubber manufacturing facilities. Waste rubber from molding operations also contribute heavily. Waste rubber can be classified into three categories: (1) uncured rubber waste; (2) cured rubber waste; and (3) off-specification products. Currently, much of the uncured rubber waste is recycled at the facility. Cured rubber waste is either recycled at the facility or sold to other companies, which use it to make products such as mud flaps and playground mats. Off-specification products can be sold to other companies, which make products from shredded or scrap rubber, or it can be disposed.

Wastewater

Contaminants are frequently added to wastewater in direct-contact cooling applications and from direct-contact steam used in vulcanizing operations. The residual containers used for dispensing contaminated and noncontaminated adhesives can also be sources of contaminated wastewater through their rinsing.

Zinc is of particular concern as a constituent of stormwater at the facilities involved in manufacturing and processing rubber products. Inadequate housekeeping is considered to be the primary source of zinc. Inefficient, overloaded, or malfunctioning dust collectors and baghouses are another source of zinc. Facilities that grind rubber usually create dust. This dust, composed partially of zinc, can go untreated (no dust collector) and be released into the atmosphere through ventilation fans. The ventilation fans, usually located in the ceilings, deposit the dust on the roofs, where it is washed by rain into stormwater drains. Some facilities use zinc stearate slurry to prevent sticking between rubber products and have indicated that the slurry frequently drips to the floor, where it may eventually drain to a stormwater outlet.

Tire Manufacturing

The resource material and pollution outputs from the tire manufacturing process include all of the outputs discussed above in rubber product manufacturing. There is, however, an emphasis on the VOC emissions, which result from cementing and spraying operations, as well as on scrap tire disposal.

Pollution Prevention Options for Plastic Manufacturing

Within the plastic industry, there are substantial pollution prevention options for most environmental concerns, including chemical spills, wastewater (including solvents in wastewater), plastic pellet loss, and plastic product disposal. Pollution prevention for leaks and spills of chemical additives during compounding or finishing operations can be as simple as covering the chemical containers as often as possible and training employees to properly handle and dispose of chemicals.

Wastewater

The pollution prevention options for process wastewater from the plastic manufacturing industry are more complex. As discussed earlier, wastewater can be divided into three categories: contact cooling and heating water; cleaning water; and finishing water. The techniques identified by EPA as appropriate for contact cooling and heating water are the activated carbon process and good housekeeping practices. EPA analysis indicates that only one pollutant of concern, BEHP, is present in contact cooling and heating water in treatable concentrations, and the only technology identified to control BEHP is the activated carbon process. To maintain low concentrations of other pollutants currently discharged in contact cooling and heating water, the application of good housekeeping practices is advised. For example, routine segregation of raw materials and lubricating oils from the cooling and heating water will keep pollutants not actually generated during the plastic molding and forming operation out of the cooling and heating water.

In cleaning water, experience indicates that there are three conventional pollutants (BOD, oil and grease, and TSS), three nonconventional pollutants (COD, TOC, and total phenols), and two priority pollutants (phenol and zinc) present in treatable concentrations. For the cleaning water category, EPA proposes pollution prevention technologies based on in-process controls. For example, one control is recycling process water through a sedimentation tank to remove the suspended solids so the process water can be reused.

Pollution Prevention Options for Rubber Manufacturing

The following pollution prevention methods have been used by various facilities:

Careful transportation mechanisms. Receive chemicals in closed docks. Use sealed containers where possible or receive bulk rail or truck shipments only from companies with a minimal history of spills. Store chemical piles inside the facility to ensure that any fugitive emissions can be contained within the facility.

Sealed containers. Provide sealed containers for all open materials. Sealed containers should have air space between the chemical and the container cover to minimize "puffing" losses when the container is opened. Similarly, placing secondary containment mechanisms around all storage containers provides further protection from spills and leaks.

Automatic dispensing. Use automatic dispensing and weighing equipment whenever possible. Automatic dispensing minimizes waste due to spills from manual dispensing and provides quality control.

Reduce toxic chemical usage. Reduce the use of toxic chemicals by reformulating the product. For example, one manufacturer claims to have reduced zinc waste simply by reducing the amount of zinc added to the compound master batch. He kept reducing the amount of zinc until the quality of the product suffered. The manufacturer then assumed that the last zinc level at which no decrease in product quality was discovered was the best "waste minimized" operating level. Since manufacturers may use several hundred compounding formulae on a regular basis, formulae review may not only reduce waste production, but also provide for increased quality control.

Computer inventories. Provide computer inventory control to minimize the amount of stock purchased.

Spills and sweeping protocols. Provide protocols for cleaning up spills and sweeping to ensure the proper segregation of waste.

Wastewater

The volume of water consumed by heating and cooling operations can be cut significantly by installing a closed-loop water cooling or heating system or a closed-loop ethylene glycol system. Another problem is that wastewater is often contaminated by oil and grease. To prevent the spilling and leaking of waste oil and grease, the following pollution prevention methods have been considered:

- Substitute lubricating grease for oil, especially for milling equipment. Grease has been shown to reduce substantially the amount of manifested waste.
- Perform preventive maintenance of processing, molding, and curing equipment. Such practices can further reduce the volume of manifested oil and grease waste by reducing waste from worn seals and gaskets.
- Recover oil from oily wastewaters prior to disposal. For instance, oily wastewaters collected from equipment engine pits could be routed through a centrally located oil/water separator prior to discharge.

Solid Waste Disposal Options

To prevent the improper disposal of scrap rubber, facilities can segregate and recycle rubber wastes. Properly segregating wastestreams may be as simple as placing a screen over part of the molding equipment so that waste rubber stock produced can be kept separate from the oily wastewaters and recycled back into the process. Other processes may include separating cured from uncured rubber, and recycling the uncured portion back into the process. Additional reduction suggestions include:

- Reclaim and recycle cured, off-specification rubber.
- Reprocess rubber by taking used rubber products and producing a form of rubber filler or an ingredient that can be incorporated into virgin

rubber compounds. The first process severs cross-links by chemical or steam digestion to produce a product known as "reclaiming." The second process is a grinding of rubber compounds by ambient grinding, cryogenic grinding, or solution grinding in water.

Scrap rubber that cannot be recycled within the manufacturing process is being manufactured into sheets and various shapes to use as athletic area surfaces and other floor coverings, sheet gasket material, and loading dock bumpers.

SIC Code 32: Stone, Clay, Glass, and Concrete Products

Although the stone, clay, glass, and concrete industry produces a wide array of products, the pollution outputs are generally limited to particulate emissions, certain solid wastes associated with raw material handling and plant maintenance, and wastewater resulting from mixing, melting, and refining raw materials and finishing the final product. Processes in this industry often entail the heating and mixing of materials in a kiln and the use of water as a cooling agent or as an ingredient in making the final product. The fuel used to operate a kiln is, itself, a source of pollution.

Environmental Impacts

The wastes generated from manufacturing structural clay products result mainly from handling raw materials, particulate emissions, plant maintenance, and pollution control equipment. Raw materials become wastes when they are spilled, off-specification, or out of date. Pollution control wastes from the clay industry include dust accumulated in baghouses and the solid residues from wet scrubbers used to treat nitrogen oxide emissions. Plant maintenance waste consists primarily of waste oil, which is generated from many types of mechanical equipment.

Similarly, waste generated in the glass industry comes from three categories of sources: materials handling, pollution control equipment, and plant maintenance. Materials handling waste consists of the waste generated during the receiving and transfer of raw materials at the facility, including raw materials that are rendered unusable when spilled during receiving or transfer.

Emission control equipment at glass manufacturing plants generates waste residues from the pollutants captured during the melting, forming, and finishing steps. The melting of raw materials to produce glass creates air emissions consisting of particulates, SO_x and NO_x, which are generated from the combustion of fuel and from the evaporation or dissociation of raw materials. Emissions are also generated during the forming and finishing of glass products as a result of thermal decomposition of lubricants.

The production of lime results in several types of pollutants. Air emissions associated with lime manufacturing include particulate matter from crushing, screening, and calcining of the limestone and combustion products from the kilns. Nitrogen oxides, carbon monoxide, and sulfur dioxide are all produced in lime kilns. Methods of emission control include wet scrubbers, baghouses, cyclones (particles forced into a cyclone-shaped vortex) and electrostatic precipitators.

Pollution outputs from cement manufacturing plants include process waste (primarily cement kiln dust), air emissions, wastewater, plant maintenance waste (such as waste oil from equipment lubrication), and research and laboratory waste. Cement kiln dust is the largest waste stream from cement plants.

Cement plants generate particulate and gaseous air emissions. Sources of particulate emissions include raw material storage, grinding and blending, clinker production, finish grinding, and packaging. The largest emission source within cement plants is the kiln operation, which includes the feed system, the fuel firing system, and the clinker cooling and hauling system. The kiln generates nitrogen oxides, sulfur oxides, carbon monoxide, and hydrocarbons as part of the normal combustion of fuel used to supply heat for cement kilns and drying operations. Cement kilns also emit trace metals and certain organic compounds.

Pollution Prevention Options

Pollution prevention techniques available to this industry can be classified into the following categories: 1) source reduction; 2) recycling and reuse; and 3) improved operating practices. Product substitution is one means of reducing paint waste generated by plants engaged in finishing of pottery products. Water-based finishes, including paints and enamels, can be substituted for solvent-based finishes, reducing the amount of volatile emissions from finishing processes. The use of water-based finishes may, however, result in hazardous waste generation and wastewater discharges. The following are some other examples:

- Increase the use of waste glass, or cullet, as a feedstock. Currently, about 67% of all cullet is landfilled or stockpiled. Glass manufacturers typically use 30% cullet along with raw materials to make new glass. Increasing the use of cullet reduces energy consumption, since it requires less energy to melt cullet than to melt other raw materials. Waste glass that is not reused on-site can be used in the production of road materials (known as glasphalt).
- Refractory scrap from glass facilities also can be recycled. Spent refractory brick can be used as a feedstock by brick manufacturers without affecting the quality of the final product.

A major quantity of hazardous waste generated from glass making is generated in the receiving and delivery areas. Improvements such as clean-up and maintenance in receiving areas can minimize this waste.

- Pave receiving areas to make collection and clean-up more efficient and effective and allow spilled material to be identified and separated for recycling back into the process.
- Keeping the receiving areas clean would allow material spills to be collected and added to the raw materials.

Air pollution control technologies used in the glass industry commonly transfer contaminants from one medium (air) to another (water or hazardous waste). Process improvements can help reduce total waste generation and improve manufacturing efficiency.

- "Rapid Melting Systems" involve preheating the batch prior to melting. This practice reduces process time, energy consumption, and air emissions.
- Substituting oxygen for combustion air can reduce nitrogen oxide and particulate emissions. The drawbacks of using pure oxygen rather than air are its higher cost and localized hot spots during combustion.

Suggestions for cement processes include the following:

- Alternative cement finishing processes, including the use of water-based and powder coatings, can reduce the amount of paint-related wastes generated by manufacturers of cement products.
- Minimize the production of cement kiln dust. There are three primary means to decrease the amount of dust generated by a kiln:
  1. Reducing gas turbulence in the kiln and avoiding excessive flow velocities

2. Using chains near the cool end of the kiln to trap the dust before it is released into the kiln exhaust
3. Using fuels with low ash contents, such as liquid wastes, and with high fuel value

- Cement kiln dust that contains alkalis or possesses other undesirable characteristics may be modified so that it can be returned to the kiln system. Modification techniques include:
  1. Pelletizing
  2. Leaching with water or a potassium chloride solution to remove alkali salts
  3. Alkali volatilization
  4. Recovery scrubbing, also known as flue gas desulfurization
  5. Fluid bed dust recovery
  6. Material accumulated from desulfurization can be ground and used as an additive to the clinker to make cement
- Cement kiln dust can be used beneficially in a variety of ways:
  1. Sold by some plants for sewage sludge solidification
  2. Used as an adsorbent for desulfurization, particularly in the cement plant's air pollution control equipment
  3. Used as a neutralization agent for acidic materials
  4. Used as a soil stabilizer
  5. Used as an ingredient in various agricultural and construction products
- Wastes generated from other industries can be used at cement kilns as fuels and raw material substitutes.
- Use wastes generated by air pollution control equipment at facilities that produce structural clay products. The waste generated by wet scrubbers can

often be returned to the production process as a raw material to replace clay or other alkaline additives.

- Waste generated during raw materials receiving can be eliminated by modifying the equipment and operating practices.

SIC CODE 33: IRON AND STEEL

Pollution Prevention Options

Raw Material Handling and Storage

The handling techniques for materials prone to dusting include the following:

- Avoid handling dust-prone material during periods of high winds.
- Minimize dust generation on roadways by using paved roadways and periodically cleaning them.
- Use dust-suppression materials on the roadways during inclement weather.
- Minimize the movement of materials.
- Design the storage piles to minimize wind erosion of material, and use dust-suppression techniques (freshwater or water collected and recycled) on the storage piles that are susceptible to dusting.
- All fixed conveyors, the railcar thaw shed, and the railcar dumper should be enclosed with dry-type dust collection at all transfer points.

Utilizing the best control technology for process wastewater, effluents can be recycled to dust suppression and car-washing operations with any surplus being directed to the blowdown treatment plant for treatment. This type of technology prevents contamination of ground water and soil and prevents contaminated water from discharging into a waterway. This can be accomplished by performing the following:

- Provide an impervious base for each storage pile with a catchment drainage channel to collect the precipitation and dust suppression liquid.

- Use a sedimentation sump to collect solids for recycling or disposal.

Formcoke

Formcoke is an alternative to conventionally produced metallurgical slot-oven coke. It is produced from a combination of carbonaceous materials and a binder. The carbonaceous solids range from low-rank subbituminous coals to high-rank, low-volatile coals together with components such as chars, coke breeze, anthracite, and petroleum coke. Binders include tars, petroleum pitches, and coal-tar pitches. The resultant mixture can be extruded or briquetted to produce uniform shapes for consumption in blast furnaces or cupolas. The briquetting step can be performed either hot or cold and may or may not be preceded by thermal treatment of one or more of the feed materials.

A significant advantage of the formcoke process is its flexibility in using raw materials. While the conventional coke-making process requires coals within a narrow spectrum of composition, virtually all types of coal, from subbituminous through anthracite, can be and have been used to produce formcoke. By combining different proportions of noncoking and coking materials, formcoke with specific properties can be produced. Furthermore, the formcoke process can be designed to accommodate virtually any feed at a single plant location. Dozens of formcoke processes have been developed to produce blast furnace coke, foundry coke, or domestic smokeless fuel; however, formcoke as a replacement for conventional blast furnace coke has been the primary target. Many processes never progressed beyond small-scale pilot plants. However, several processes have progressed to near-commercial-scale production plants and successful demonstration of the product in blast furnaces or cupolas.

Sintering Plants

The recommended pollution control criterion is zero discharge for sinter plants. All solid waste should be recycled in the plant. However, where concentrations of alkalis build up in the collected dust, it may be necessary to dispose of these materials elsewhere. One alternative practiced in Japan is to sell this material to cement plants. Currently, processes are being developed for the direct use of ore, with no sinter plant required. However, if a sintering plant exists within the facility, the following list provides options to help prevent pollution:

- Modify the on-strand cooling to increase heat recuperation, sinter quality, and dust collection.
- Increase the quantity and variety of steel plant, iron-bearing materials that are sintered. This is especially true for iron-bearing metals containing low-temperature melting metal oxides and alkali materials with low melting points.
- Utilize "dry" air pollution control systems, which eliminate any water discharge from the air pollution control systems.

Blast Furnace

- Use of beneficiated charge materials: The quality of the blast furnace raw materials is leading increasingly to greater beneficiation of the iron-bearing materials and especially the greater use of pellets. This trend, together with the increased addition of fluxes to the pellets (or sinter), has increased blast furnace production due to the reduction in slag volume and improved furnace operation.
- Use of high hot-blast temperatures: Improved stove-firing techniques, improved stove-changing

equipment, the use of higher-quality raw materials (beneficiated charge materials), and the improved control of blast moisture have resulted in increased fuel efficiency. These developments have led to the use of higher hot-blast temperatures, which in turn has resulted in improved furnace operation, higher furnace production, and increased productivity. Hot-blast temperatures of 1200 to 1250°C are now common, and temperatures as high as 1272°C are being achieved.

- Oxygen enrichment of the blast: New techniques for controlling the flame temperature, together with blast moisture control and fuel injection, have led to increased enrichment of the blast air. It has been proven that each percent increase in oxygen in the blast air can increase the production rate by 2% to 4% and significantly increase fuel efficiency.
- Utilize computer controls, especially with newly developed sensors, to assist the operators in maintaining a smooth-running furnace operation.
- Emissions from the slag in the cast house should be collected with the emissions from the iron taping and handling. One control technology for the reduction of total reduced sulfur (TRS) and other emissions from the slag quenching and cooling is the use of a pelletizing or granulating technology in an enclosed space. The plume from the process should be ducted to a wet scrubber to minimize the particulate and TRS emissions.
- Pelletizing or granulation technology also reduces the emissions from slag crushing, screening, and handling since the product does not require primary crushing and is relatively dust-free.
- Sell the blast furnace slag as a construction aggregate, railroad ballast or, when pelletized or granulated, as an aggregate for low-weight

concrete. Other uses include mineral wool, glass manufacturing, sewage treatment, roofing, and soil conditioning.
- Iron-bearing material collected by the blast furnace gas-cleaning system should be recycled wherever feasible. In some cases the material may be recycled through a sinter plant. An emerging technology is cold bonding briquetting for use as part of the blast furnace burden. (If this material is high in zinc or lead, further processing may be necessary before recycling to the blast furnace.)

Hot Metal Desulfurization

A high-efficiency wet scrubber, electrostatic precipitator, or fabric filter with effective hooding over the ladle car at the desulfurization station is the best technology for controlling emissions from the hot metal desulfurization. Since this is an intermittent process, the air pollution control system for the cast house is often used for this application. If the hot metal desulfurization is carried out at the steel-making shop, the ladle transfer dust collection system is usually used.

Basic Oxygen Furnace

In the interim the basic oxygen process will be further refined with the objectives of increasing quality and yields and reducing costs and environmental impact.

- Reduce slopping by improving process control and operating practices so that the hood can be moved closer to the converter vessel.
- Basic oxygen slag should be processed for steel recovery, partial recycle to blast furnace and sale of the balance as a by-product. Some basic oxygen furnace slag is used as aggregate in asphalt for road paving. Other uses include ice control and soil conditioning.

Electric Arc Furnaces

The increased computerization of electric arc furnace melt shops will continue with developments being made in the field of computer programs and measuring devices. The objectives are to improve furnace operations and steel quality, reduce the consumption of materials and electric power, increase work force productivity, and reduce overall waste.

- Electric furnace slag should be processed for steel recovery and resale as a by-product. The crushed slag is often used as a construction aggregate.
- Furnace dust should be recycled where feasible. Note that electric furnace dust often has a high zinc concentration and should be processed in a Waelz kiln or similar process with zinc oxide recovery before the dust is suitable for recycling.
- Recover zinc from galvanized steel prior to charging as scrap to an electric arc furnace.
- Use an induction-heated holding furnace following the electric arc furnace to feed the metal in a pseudo-plastic state to multiple continuous casting machines.

Semifinished Product Preparation

There are a number of benefits derived from continuous casting that directly or indirectly result in reduced pollution including:

- Increased yields from liquid steel from the ladle to the semifinished rolled shape results in a reduction in scrap and scale losses in three areas: the primary rolling mill, the pouring operation (primarily elimination of top and bottom ingot crop losses), and the soaking pits (primarily scale losses). The process yield is more than 95% for continuous casting compared with approximately 80% for the production of semifinished products from ingots.

- Improved quality: An improved, more uniform finished product can be obtained with fewer internal and surface defects.
- Energy savings are achieved with continuous casting because energy-consuming steps in the ingot process are eliminated. Also, a large proportion of the energy contained in the molten steel is retained in the semifinished product, thus requiring less energy for reheating prior to rolling.
- Continuous casting reduces pollution through the elimination of ingot processing facilities such as soaking pits and the primary rolling step.
- Productivity is improved through the reduction in manpower and the shortening of the elapsed time required to produce semifinished steel from liquid steel.

SIC CODE 34: FABRICATED METAL PRODUCTS

Environmental Impacts

Each of the metal shaping processes can result in wastes containing chemicals of concern. For example, the application of solvents to metal and machinery results in air emissions. Additionally, wastewater (containing acidic or alkaline wastes and waste oils), solid wastes (such as metals), and solvents are usually generated during this process. Metal fabrication facilities are major users of solvents for degreasing. In cases where solvents are used solely in degreasing (not used in any other plant operations), records of the amount and frequency of purchases provide enough information to estimate emission rates, based on the assumption that all solvent purchased is eventually emitted.

Metalworking fluids are applied to either the tool or the metal being tooled to facilitate the shaping operation. Metalworking fluid is used to:

- Control and reduce the temperature of tools and aid lubrication
- Control and reduce the temperature of workpieces and aid lubrication
- Provide a good finish
- Wash away chips and metal debris
- Inhibit corrosion or surface oxidation

Fluids resulting from this process typically become contaminated upon extended use and reuse. In general, metalworking fluids can be petroleum-based, oil-water emulsions, and synthetic emulsions. When disposed, these fluids may contain high levels of metals (e.g., iron, aluminum, and copper). Additional contaminants present in fluids resulting from these processes include acids and alkalis (e.g., hydrochloric, sulfuric, nitric acids), waste oils, and solvent wastes.

Scrap metal may consist of metal removed from the original piece, and may be combined with small amounts of metalworking fluids (e.g., solvents) used prior to and during the metal shaping operation that generates the scrap. Quite often, this scrap is reintroduced into the process as a feedstock. The scrap and metalworking fluids, however, should be tracked since they may be regulated as solid wastes.

The following suggestions cover most facets of the metal fabrication industry.

Metal Shaping Operations

- Use specific pipes and lines for each set of metals or processes that require a specific oil to reduce the amount of cleanouts.
- Extend machine coolant life by centrifuging and adding biocides.
- Install a second high-speed centrifuge on a system already operating with a single centrifuge to improve recovery efficiency even more.
- Install a chip wringer to recover excess coolant from aluminum chips.
- Use a coolant analyzer to allow better control of coolant quality.
- Use an ultrafiltration system to recover soluble oils from wastewater streams.
- Substitute a hot lime bath or borax soap for oil in cold forming or other processes where oil is used only as a lubricant.
- Use a stamping lubricant that can remain on the piece until the annealing process, where it is burned

off. (This eliminates the need for hazardous degreasing solvents and alkali cleaners.)

- If filtration or reclamation of oil is required before reuse, segregate the used oils to prevent their mixing.
- Segregate metal dust or scrap by type for resale (e.g., sell metallic dust to a zinc smelter instead of disposing of it in a landfill).
- Oil-scrap mixtures can be centrifuged to recover the bulk of the oil for reuse.
- Perform on-site purification of hydraulic oils using commercial "off-the-shelf" cartridge filter systems to extend the service life of the oil.
- Use a settling tank (to remove solids) and a coalescing unit (to remove tramp oils) to recover metalworking fluids.

Solvent Cleaning

- Use countercurrent solvent cleaning (i.e., rinse initially in previously used solvent and progress to new, clean solvent).
- Cold clean with a recycled mineral spirits stream to remove the bulk of oil before final vapor degreasing.
- Reduce grease accumulation caused by excess oil applications by adding automatic oilers.
- Use plastic blast media for paint stripping rather than conventional solvent stripping techniques.
- Use less-hazardous degreasing agents, such as petroleum solvents or alkali washes. For example, replace halogenated solvents (e.g., trichloroethylene) with liquid alkali cleaning compounds. (Note that compatibility of aqueous cleaners with wastewater treatment systems should be ensured.)

- Substitute chromic acid cleaner with nonfuming cleaners such as sulfuric acid and hydrogen peroxide.
- Substitute less-polluting cleaners such as trisodium phosphate or ammonia for cyanide cleaners.
- Recycle spent degreasing solvents on-site using batch stills.
- Arrange a cooperative agreement with other small companies to recycle solvent centrally.

Chemical Treatment of Fabricated Metal Parts

- Increase the number of rinses after each process bath and keep the rinsing countercurrent to reduce drag-out losses.
- Recover unmixed acids in the wastewater by evaporation.
- Reduce rinse contamination via drag-out by:
  1. Slowing and smoothing removal of parts
  2. Rotating parts, if necessary, using surfactants and other wetting agents
  3. Maximizing drip time and using drainage boards to direct dripping solutions back to process tanks
  4. Installing drag-out recovery tanks to capture dripping solutions
  5. Using a fog spray rinsing technique above process tanks
  6. Using techniques such as air knives or squeegees to wipe bath solutions
  7. Changing bath temperature or concentrations to reduce the solution surface tension
- Clean copper sheeting mechanically with a rotating brush machine that scrubs with pumice.
- Change copper bright-dipping process from a cyanide dip and chromic acid dip to a sulfuric acid/

hydrogen peroxide dip. (The new bath is less toxic and copper can be recovered).

- Recover copper from brass bright-dipping solutions using a commercially available ion exchange system.
- Replace barium and cyanide salt heat treating with a carbonate–chloride carbon mixture, or with furnace heat treating.
- Sell waste pickling acids as feedstock for fertilizer manufacture or for neutralization and precipitation.
- Recover metals from solutions for resale.
- Send used copper pickling baths to a continuous electrolysis process for regeneration and copper recovery.

Plating Operations

- Implement counter flow rinsing and cascade rinsing systems to conserve water.
- Use drip bars to reduce drag-out.
- Use drain boards between tanks to capture drag-out.
- Install racking to reduce generation of drag-out.
- Regenerate plating bath by activated carbon filtration to remove built-up organic contaminants.
- Install pH controller to moderate the alkaline and acid concentrations in tanks.
- Install atmospheric evaporator to reduce metal concentrations.
- Use reactive rinsing and multiple drag-out baths.
- Improve control of water level in rinse tanks, improve sludge separation, and enhance recycling of supernatant to the process by aerating the sludge.
- Substitute sodium bisulfite and sulfuric acid for ferrous sulfate to oxidize chromic acid wastes, and substitute gaseous chlorine for liquid chlorine to diminish cyanide reduction.

- Recover process chemicals by fog rinsing the parts over the plating bath.
- Use ion exchange and electrowinning, reverse osmosis, and thermal bonding when possible.
- Use sludge slagging techniques to extract and recycle metals.
- Use hydrometallurgical processes to extract metals from sludge.
- Convert sludge to smelter feed.
- Remove and recover lead and tin from boards by electrolysis or chemical precipitation.
- Implement the electrodialysis reversal process for metal salts in wastewater.
- Oxidize cyanide and remove metallic copper to reduce metal concentrations.

Finishing Operations

- Use an airless gun for heavy film application.
- Use an air-assisted spray gun for a wide range of fluid output.
- Preinspect parts to prevent painting of obvious rejects.
- Ensure the spray gun air supply is free of water, oil, and dirt.
- Change from conventional air spray to an electrostatic finishing system.
- Regenerate anodizing and alkaline silking baths with contemporary recuperation of aluminum salts.
- Segregate nonhazardous paint solids from hazardous paint solvents and thinners.
- Recycle metal sludges through metal recovery vendors.
- Use membrane filtration or use activated carbon to recover solvent vapors.

Paint Cleanup

- Reduce equipment cleaning by painting with lighter colors before darker ones.
- Reuse cleaning solvents for the same resin system by first allowing solids to settle out of solution.
- Flush equipment first with dirty solvent before final cleaning with virgin solvent.
- Use pressurized air mixed with a mist of solvent to clean equipment.
- To prevent spray gun leakage, submerge only the front end (or fluid control) of the gun into the cleaning solvent.
- Solvent wastestreams should be kept segregated and free from water contamination.
- Solvent recovery units can be used to recycle spent solvents generated in flushing operations.
- Install a recovery system for solvents contained in air emissions.
- Use a small solvent recovery still to recover spent paint thinner from spray gun clean-ups and excess paint batches.

4

Pollution Prevention in Relation to Processes and Hazardous Material Function

SOLVENT CLEANING AND DEGREASING

Solvent cleaning and degreasing is the process of using an organic solvent to remove unwanted grease, oils, and other organic films from surfaces. Solvents are used for cleaning, degreasing, and rinsing in vehicle maintenance operations; in the manufacture of aircraft, motor vehicles, metal furniture, tools, appliances, paint, ink, organic and agricultural chemicals, adhesives, petrochemicals, plastics, rubber, vegetable oil, textiles, and other products; and in photographic processing, coating, printing, dry cleaning, and their industries. In metal finishing industries, for example, cleaning usually follows machining and precedes other surface-finishing steps. During machining, a variety of oil emulsions or synthetic fluids are sprayed on the workpiece (parts being fabricated) for lubrication or cooling. These fluids must be removed before final finishing. The slightest amount of oil or other residue can render the final surface finish ineffective. Pollutants generated from degreasing include liquid waste solvent compounds containing unwanted film material and air emissions containing volatile solvents.

Process Description

Vapor degreasing is a cleaning practice common to many industries. It involves using nonflammable, chlorinated hydrocarbons in a vapor phase to clean metallic and other suitable surfaces. The special apparatus—the degreaser—that provides solvent vapor consists of a tank that is one-tenth full of solvent. The solvent is heated, usually by steam coils, to its boiling point, producing solvent-saturated vapor in the upper portion of the tank. The item to be cleaned is inserted either manually or automatically into the vapor region. Hot solvent vapor immediately condenses onto the surface of the item. The condensed solvent then drips back into the liquid bath and takes with it the removed dirt and grease. Solvent vapor is prevented from escaping to the atmosphere by use of a refrigerated section on the upper part of the tank. The solvent condenses on the walls of this section and returns to the liquid sump. Cold cleaning and spot cleaning are two other solvent cleaning techniques. In cold cleaning, the part to be cleaned is immersed in a tank of solvent and typically agitated mechanically or ultrasonically. In this instance, the cleaning performance degrades with use because the solvent bath becomes loaded with contaminants—a distinct disadvantage as compared to vapor degreasing, because solvent vapors remain relatively pure. The typical method of spot cleaning is to use a lint-free cloth saturated in solvent to clean a specific area or a part or workpiece, often as a part of a repair operation.

Vapor degreasing and other solvent cleaning frequently require the use of chlorinated solvents. Chlorinated solvents were adapted by industry because of their ability to dissolve a wide range of organic contaminants, their low flammability, and their high vapor pressure, which allows them to evaporate at room temperature and leave a residue-free surface. Prior to 1995, popular chlorinated solvents included 1,1,2-trichloro-1,2,2-trifluoroethane (CFC-113); 1,1,1-trichloroethane (TCA, also called methyl chloroform); trichloroethylene; tetrachloroethylene (also called perchloroethylene, or PERC); and dichloromethane (or METH for methylene chloride). Concerns about the toxicity and environmental effects (ozone depletion in the upper atmosphere) of continued use of chlorinated solvents have led chemical companies to search for replacement solvents. Pollution prevention efforts focus on reducing or eliminating the use of these solvents.

Pollution Prevention Options

Rethinking Cleaning

When attempting to reduce the use of hazardous cleaning solvents, a facility should first fully understand the cleaning process being practiced. Does the part in question need to be cleaned with a chlorinated solvent, or even need to be cleaned at all? Can modifications be made to a process prior to the cleaning step that could obviate the cleaning step altogether? Once it is established that cleaning is unavoidable, alternative cleaning methods need to be considered.

Aqueous Cleaning

Cleaning using water-based solutions and various detergents ("aqueous cleaning") is an alternative to conventional chlorinated solvent cleaning. Aqueous cleaning systems can include alkaline solutions and detergents to enhance their soil removal capabilities, or consist solely of hot-water washes. Water-based cleaners have few or no volatile components, which means cleaning cannot take place in the vapor phase. Immersion tanks are most commonly used in these applications. These systems are often used with pressurized sprays, agitation, filtration, heat, or some other physical process to further provide effective cleaning. Agitation can be accomplished either with ultrasonics or by mechanically rotating the part, and/or circulating the system. An aqueous system usually requires many tanks (often more floor space than vapor degreasers), and usually requires adding rinsing and drying steps after cleaning to accomplish comparable solvent degreasing results.

Ultrasonics is a process that uses high-frequency sound waves to improve the efficiency of aqueous and semi-aqueous cleaners. By generating zones of high and low pressures in the liquid, the sound waves create microscopic vacuum bubbles that implode as the sound wave moves and the zone changes from negative to positive pressure. This process, called cavitation, exerts enormous localized pressures (~10,000 psi) and temperatures (~20,000 F) on a microscopic scale that loosen contaminants and actually scrub the workpiece. Ultrasonic energy is applied to a solution by means of a transducer, which converts electrical energy into mechanical energy. Using ultrasonics as part of an aqueous system can be particularly good at reaching into crevices and small holes to clean.

Unlike some chlorinated solvents, which effectively clean almost any surface, an aqueous cleaning system usually needs to be specifically tailored for particular cleaning surfaces. For this reason, converting to an aqueous system can be expensive in the short term, but very cost-effective in the long run. In some cases, cleaning suppliers will formulate cleaners for the client. The combination of additives chosen for an aqueous cleaning system alters the foaming, wetting, and soil-removal properties of the solution. Different additives include synthetic detergents and organic surfactants, saponifiers, acids, alkalines, and corrosion inhibitors. Some alkaline cleaners, due to their high pH, need to be tested for potential corrosion characteristics prior to use on sensitive materials. A rinsing step is required to remove residual films left by these additives in the cleaning process.

Reusing and Recycling

Solvents can be reused by recycling and distillation. Frequently, high-quality solvents are used once for high-precision cleaning and are then disposed of, either on a scheduled basis or because they do not meet their original specifications. These solvents can be reused for applications that do not require such purity ("cascade" reuse). Technologies are also available for cleaning solvents so that they can be reused for their original purpose. Distillation simply involves heating a solvent until it vaporizes, then condensing the vapor. The condensed vapor is then reused. Generally, on-site distillation equipment is economically favored over commercial off-site recovery. Off-site recovery incurs costs such as transportation, the risk of spills, and the liability of improper disposal of still bottoms. Commercial systems today recover 70–90% of solvent input.

Vapor Degreaser Modifications

Conventional open-top vapor degreasers allow for quick and thorough cleaning, but produce significant air emissions and solvent losses. Emissions from vapor degreasers are specifically due to workpiece entry and removal, dragout, idling, and degreaser shutdown. The following tips can cut down emissions from traditional open-top vapor degreasers:

- Be sure to place a cover on the tank opening during idling and shutdown.
- Increase the freeboard height (the height of the tank walls above the vapor layer) to better contain the solvent vapors.
- Convert water-cooled coils to refrigerated coils to better contain solvent vapors.
- Lower and raise workpieces gently to avoid turbulence.
- Retain workpieces over the tank for a short while to cut down on drag-out.

SURFACE COATING AND PAINTING

Surface coatings are employed wherever it is desired to provide decoration, protection, and/or safety marking to a product or item. Most coatings applied to metallic substrates are presently solvent-based, although replacements offer significant VOC reductions. Most metallic substrates undergo stripping, degreasing, and cleaning before a product coating can be applied, since the surface must be free from contamination. The purpose of surface treatment is to condition or prepare the surface so that the paint forms a better bond with the metal.

Process Description

Numerous methods for surface coating exist. For a large number of very small items, the most commonly used methods are tumbling, barreling, or centrifuging. For cylindrical items, a commonly used method is dipping. Flow coating is often employed for items that would be difficult to dip because of their size or shape, or as a means of avoiding the installation and operation of large dip tanks. The most widely used method for coating application is still the spray gun.

A spray gun operates by using compressed air to atomize the paint and produce a fan or circular cone spray pattern. Some of the variations on spray gun painting are airless spray guns, electrostatic spray guns, and High-Volume-Low-Pressure (HVLP) spray guns. Airless spray guns force the paint out at high pressure so that no air is required for atomization. Electrostatic spray units are designed so that the atomized paint leaving the gun has a positive charge, which causes the paint to be attracted to the grounded object.

The primary wastes associated with product coating applications consist of VOC-laden solvents, spent cleaning solutions, spent stripping solutions, empty containers, and equipment cleaning wastes.

Pollution Prevention Options

Coating Application Waste

The wastes generated during coating application are primarily due to: (1) paint overspray, and (2) the failure of all of the paint to reach the target. Minimizing the overspray and maximizing the paint that reaches the product can be achieved by selecting proper application equipment.

Overspray can be reduced from an estimated 50% with an airless gun to about 15% with the HVLP gun. The coating materials used are heated to reduce viscosity, which allows for use of lower ejection pressures.

Spray gun cleaning is a major source of waste generation. The spray guns are usually rinsed with solvents after use to remove all painting residues. Two approaches to reducing cleaning waste currently exist. First, disposable plastic liners, available for most spray gun paint reservoirs, effectively reduce the amount of solvent waste generated by reducing paint-to-equipment contact. Second, closed-system spray gun cleaners are now available at a relatively low cost ($500). Such gun cleaners avoid the discharge of solvent into the air while cleaning a spray gun and use a minimum of solvent.

Operator Training and Maintenance

Since many spray systems are manually operated, the equipment operator has a major impact on the amount of waste produced. When air pressures are set too high, the paint has a tendency to bounce off the surface and increase overspray. Keeping it perpendicular to the surface instead of arching the spray gun decreases the overspray and produces a more even coating. Experimentation with the speed, distance, and angle of coating will be necessary to minimize overspray with specific equipment.

Whenever a defective finish is produced it needs to be reworked, producing waste and reducing profits. This includes restripping and cleaning, as well as coating. Preventive maintenance plays a critical role in reducing the amount of overspray, stripping waste, and equipment cleaning waste produced. Maintaining all application equipment in good working order will decrease the likelihood of producing a bad coating. Spray guns should be cleaned after use or whenever there will be an interval between uses. Automatic paint equipment cleaning systems will accomplish this and release less solvent into the air. By fully inspecting parts before they are painted, the coating of potential rejects can be avoided. Discovering potential rejects before painting can preempt waste generation.

Material Substitution

Conventional solvent-based systems for coating application often pose potential health hazards due to the emission of toxic solvents and the disposal of a large amount of waste sludge. High solids coatings have helped to reduce the problem. To minimize the quantity and toxicity of waste paint requiring disposal, new coating types have been developed that eliminate the hazardous components of the paint and also allow for maximum reuse of paint overspray, thus increasing the efficiency of the paint application process. Alternatives include the use of water-based, radiation-curable, or powder-coating techniques.

Water-based coatings are paints containing a substantial amount of water (often as much as 80%) in place of the volatile solvent. One feature that makes water-based coatings attractive is that no major equipment changes are necessary to apply water-based coatings with solvent-based coating equipment. Another advantage is in the ease of recovering paint overspray. Overspray from a water-based coating can be collected or captured with water in the spray booth. The solution can subsequently be concentrated and reused again. Water-based coatings can also provide energy savings by making possible the recirculation of the hot air used to cure the paint.

Another alternative to solvent-based coatings is radiation-curable coatings. These coatings do not contain or use organic solvents. Reactive monomers are applied as a liquid to a surface that is then exposed to high-energy radiation such as ultraviolet (UV) or electron beam (EB). The advantages of using this coating technique include the reduction in waste from solvent loss and a decrease in energy and maintenance requirements.

Powder coatings represent an attractive third alternative to solvent-based coatings. With manual systems, the powder is sprayed on the object and the overspray is readily retrieved and recycled, something that is very difficult to do with liquid paints. Overspray and other unused powder are returned to the feed hopper for reuse. This ability to recycle the coating material provides a very high efficiency-to-use ratio. In a well-designed spray system, the coating powder remains clean at all times and the potential for waste generation (contaminated powder) is nearly zero. However, the use of dry powder is limited to work pieces able to withstand typical curing temperatures.

Where effective substitution is not possible, recycling the solvent-laden cleaning and overspray mixtures will reduce overall VOC emissions. This can be accomplished by two methods: collecting and reusing the mixtures in the next compatible coating formulation, and separating and reusing solvents from sludge filtration and centrifugation.

Binding and Amalgamation

Binding materials are used in numerous instances to amalgamate, compose, and reinforce various substances. For example, they are used in the manufacturing of building materials (including plywood), roofing board, and composite boards. They are also applied to paper pulp to strengthen construction resins, wall boards, and book bindings. They can also easily be applied to synthetic plastics and polymers to produce a multitude of manufactured products such as fiberglass insulation, automobile components, boat hulls and bodies, building panels, appliances, and sporting equipment. Binding adhesives provide reinforcing, structural strength, and rigidity, as well as heat resistance. Binders are also used in the textile industry to produce a desired appearance and stiffness. Some finishings involving binders are fabric treatments with oil repellents, fire retardants, lamination polymers, and soil release agents.

Typically, bound coatings fall into two main categories: passive coatings and active coatings. Bound coatings simply cover a surface and act as a barrier against corrosion, rotting, or weathering, such as wood preservatives. Conversely, active coatings contain chemical agents or additives, such as fungicides, to specifically reduce or eliminate the deterioration process.

Process Description

Almost as numerous as the uses for binding materials are the industrial processes used to apply them to the specific substrates. Dip tanks are commonly used for larger applications such as lumber and wood products. Of course, binders may be dispensed using one of several spray gun processes. Pressurized spray systems are mostly used for soft wood while roller applications are common to hard wood. Brush application of binders serves as a nonpressure method of application. The process known as lay-up applies sheets of veneer as adhesives to materials that are later assembled by presses into plywood.

Fiberglass production involves the combination of a polymerizing resin and a reinforcing material either sprayed onto a mold or coated onto the fibers. A liquid binder is injected through jets at the bottom of a trough to provide resilience and strength. The fiber is subsequently cured by the steam as the materials enter an oven.

Production processes in the textile industry mostly involve wet processing, which varies with the fabric. The production line is usually a continuous process that produces little to no wastewaters. Textile coatings consist of a uniform application of an elastomeric, thermoplastic polymer solution, a vinyl plastisol, or organosol across at least one side of the fabric surface or substrate. Coatings are applied by blade, roll coaters, reverse roll coaters, or possibly by rotogravure coaters. Products like rainwear and tents are commonly coated fabrics.

Dip processes are commonly used and offer the best control for dripping loss of binders. Light coatings are applied in an on-machine process where productions are limited by coatings and grades of coatings. Heavy coatings are applied in an off-machine process where production allows more flexibility in types and grades of coatings.

Binders are based mostly on phenol formaldehyde resins and melamine resins for bonding, silicone compounds for water repellence, soluble or emulsified oils for lubrication, wetting agents for control of surface tension, and extenders or stabilizers. Fire retardant additives are usually reacted in the resin formulation. Typical hazardous wastes include solvated resin, diluted resin, and partially cured resin from overspray or unused resin. Oil-borne creosote and pentachlorophenol are commonly used as wood preservatives. Binder dryings tend to release organic aerosols, gaseous organic compounds, and small amounts of material fibers to the atmosphere. Phenol formaldehyde for softwood and urea formaldehyde for hardwood are the most common adhesions made from resins and used for wood preservatives.

Reconstituted wood products are composed of raw wood combined with resins and pressed into a board. Air emissions from dryers result in VOC wastes. Emissions from presses depend on the type of resin used to bind the wood together. Hot presses result in condensable organic vapors including resin ingredients containing formaldehyde, phenol, and MDI. Phenol formaldehyde for softwood and urea formaldehyde for hardwood are the most common adhesions made from resins and used for wood preservatives. Other VOCs frequently emitted from binding processes are acetone, ammonia, methanol, methyl ethyl ketone, methyl isobutyl ketone, toluene, and xylene (mixed isomers).

Pollution Prevention Options

Wastes can be reduced by material substitution or by minimizing the number of hazardous materials used. Reuse and recovery should be incorporated to further reduce pollution.

Material Substitution

- Replace toxic chemical treatment with a nontoxic type of treatment to achieve surface protection.
- Use alternative adhesives of less-toxic formulations in panel protection—MDI substitution in the oriented strand board industry.
- Use high-moisture bonding adhesives that eliminate some drying needs and drier emissions to save energy.
- Use naturally derived adhesives (furfuryl alcohol resins, lignin adhesives, or polyvinyl acetate) to reduce hazardous air pollution.
- Source reduction in adhesive manufacturing is driven by industrial customer use. Requiring adhesives that will not emit VOC emissions like solventless water based, 100% solids, and hot melt products will result in lower hazardous waste industrywide.

Waste Reduction

- Use high-velocity spray systems that generate fewer process residuals and less drippage.
- Use drainage collection devices on roof tops to prevent rainwater from mixing with process waste.
- Use foam-extrusion to apply adhesives to veneer for a reduction of overspray to lessen wastes and dryout before pressing.
- Use water-borne products to reduce oil-borne wastes in wood preserving.
- Clean the wood products prior to treatment with biodegradable soap and remove dust with air overpressure to reduce accumulation of debris in treatment systems.
- Good housekeeping and inspection of equipment on a regular basis may reduce leaks and maintenance, and minimize waste and wear on equipment.
- Segregate wastestream.
- Improve recycling by segregating solvent wastes and water solvents, control concentrations, and label wastes.
- Overspray can be eliminated or reduced through spray reorientation activities and equipment modifications.
- Incorporate nonspray resin application methods (depending on product) to eliminate overspray.
- Dry out and dispose of excess water-based adhesives as solid waste to landfills or recycled and reused.
- Install an overhead conveyor system to eliminate production delays from drying times so that freshly glued products will travel on the system for added drying time.

- Modify the use of adhesives to maximize the use of water-based (nonhazardous) glues. Spot glue 10% of the surface area with a quick-drying solvent-based adhesive and 90% with a slower-drying water-based adhesive.
- Install an automatic adhesive–solvent mixing system for solvent-based adhesives. The amount of unused adhesives and solvents will be reduced.

ACIDS AND BASES

Thus far, focus has been directed toward pollution prevention in relation to various industrial processes and it will continue as such in the remaining sections. Basically the format has been one in which a process is identified along with its application in a number of industries and applicable pollution prevention measures that can be taken. However, this particular section deviates slightly from the aforementioned format in that it concentrates more on the function of hazardous materials, specifically acids and bases in a number of industries, along with applicable pollution prevention technologies.

The Resource Conservation and Recovery Act of 1976 defined hazardous waste and authorized the United States Environmental Protection Agency to characterize such wastes so that they could be identified. One such characteristic is corrosivity. If a representative sample of the waste is aqueous and has a pH less than or equal to 2 or greater than or equal to 12.5, then it is considered hazardous. Although this definition applies to waste materials, it provides a measurable concept of what characterizes any material as corrosive. Further, a corrosive material may destroy body tissues, metals, plastics, and other materials upon contact. The corrosiveness of acids and bases can be compared on the basis of their ability to dissociate, or form ions, in solution. Those that form the greatest number of hydrogen ions (H^+) are the strongest acids, while those that form the most hydroxide ions (OH^-) are the strongest bases. The pH, or hydrogen ion exponent, is a measure of the hydrogen ion concentration of a solution. On the pH scale of 0 to 14, low pH values indicate acidity and high values indicate alkalinity.

Acids and bases have many functions in industry. They are common reactants in the manufacture of inorganic chemicals. An example is the use of hydrochloric acid in the dissolution of minerals in refining operations. They are also commonly used to strip, clean, and prepare metal surfaces for plating, etching, and coating operations. The following sections present noteworthy industry-specific uses of acids and bases along with feasible pollution prevention alternatives and applicable pollution control technologies.

Process Descriptions

Steel Industry

In the steel industry (Standard Industrial Code 331) surface quality is an important consideration in the preparation of a finished product. In the case of steel, the surface must be prepared so that protective coatings, applied to prevent corrosion, will properly adhere. Coating processes include galvanizing (zinc coating), tin and chromium coating via electroplating, aluminizing, and terne coating (lead and tin). Prior to the application of these coatings, mill scale, rust, oxides formed during the hot-rolling forming method, grease, and soil all have to be chemically removed from the surface of the steel using solvent cleaners, pressurized water or air blasting, cleaning with abrasives, alkaline agents, or acid pickling.

In the acid pickling process, the steel surface is chemically cleaned of scale, rust, and other materials with inorganic acids such as hydrochloric, nitric, and hydrofluoric acids. After pickling, steel is generally passed through a series of rinses. Alkaline cleaners are involved, generally prior to cold rolling, in a rinsing series for the purpose of removing mineral oils, animal fats, and oils. Some common alkaline cleaning agents include caustic soda, alkaline silicates, and phosphates.

Regarding waste treatment, the primary concern is treating spent acid pickle liquor and alkaline cleaning agents. The EPA developed a model treatment process for the acid pickling operation's wastewater. The process consists of spent acid neutralization, fume scrubber recycling, cascade rinsing, equalization of all wastewaters, oil separation, lime precipitation/neutralization, flocculation with polymer, aeration, clarification, and vacuum filtration of the clarified underflow. These processes are to be done prior to hydrochloric acid regeneration.

The EPA also developed a model treatment process, based on a high level of recycle, for effluent regulation development for alkaline-cleaning operation wastewater. The process consists of equalization with oil skimming, neutralization with acid, and flocculation with polymer. Solids can then be removed by clarification followed by vacuum filtration for dewatering the sludge collected in the clarifier.

Metal Mining Industry

In this industry, SIC 10, ore beneficiation involves processing to regulate product size, the removal of unwanted constituents, or improving the quality, purity, or grade of the desired product. One of the common beneficiation steps is leaching. Leaching is the process of extracting a soluble metallic compound from an ore by selectively dissolving it in a solvent such as water, sulfuric acid, or hydrochloric acid. The desired metal is then removed from the leach solution by chemical precipitation or an electrochemical process.

For example, the beneficiation of lead and zinc ores includes crushing and grinding: filtration, sizing, flotation, and sintering of concentrates. After crushing and grinding, or milling, common conditioners inclusive of lime, soda ash, caustic soda, and sulfuric acid are used to treat the ore. The ore is then slurried in fresh or salt water. Several separate flotation steps can follow in order to concentrate individual metal values from the ore. Reagents used in the flotation processes typically include such chemicals as sulfur dioxide, zinc sulfate, coal tar, copper sulfate, and sodium or calcium cyanide.

In the case of copper extraction, acidic leaching agents can be recovered after extraction. However, depending on the metal being extracted, acid leaching can produce large volumes of metal-bearing acid solutions that need to be properly contained in leach dumps and associated extraction areas. Potential problems include the seepage of acid solutions through soils and liners beneath leach piles, leakage from solution-holding ponds and transfer channels, spills from ruptured pipes and recovery equipment, pond overflow caused by excessive runoff, and ruptures of dams or liners in solution holding ponds. The design and construction of leach pads and holding ponds have improved over the years to prevent products and reagents from seeping into the environment.

Metal Finishing Industry

The electroplating, polishing, and anodizing industry is classified under the Standard Industrial Code 3471. In such industries, acids and caustics find their primary uses in surface preparation prior to application of a surface treatment. To remove greases and soils, the operator would typically use detergents, solvents, caustics, and other media, followed by a rinse. Then the workpiece would proceed to an acid dip for the removal of oxides, followed by a rinse to remove the acid.

There are acid recycling systems available that can be used for the recovery of nitric, hydrochloric, hydrofluoric, and sulfuric acid from concentrated baths. One such technology is diffusion dialysis. In utilizing such an operation, acid is pumped through the system so as to come in contact with one side of an anionic exchange membrane. Water can then be pumped countercurrently along the recovery side of the membrane. A major portion of the acid migrates through the selective membrane, leaving metallic contaminants behind. Purified acid can then be directed back to the process tank, leaving a contaminant-laden spent acid stream ready for further processing in a metal recovery or waste treatment system.

Electronics and Computer Industry

In this industry, SIC 36, acids play a significant role in etching. Etching is a method of cutting into, or imprinting on, the surface of a material. The materials of concern are semiconductors, integrated circuits, and printing wiring boards. Acids such as sulfuric, phosphoric, nitric, hydrofluoric, and hydrochloric are used primarily in wet etching processes.

An exemplary etching process is that utilized in the manufacturing of printed wiring boards (PWBs). Printed wiring boards are the physical structures on which electronic components such as semiconductors and capacitors are mounted. PWBs are prepared utilizing photographic tools. Such tools transfer the circuit pattern to the two-sided, etched dielectric boards. The boards are then cleaned in preparation for electroless plating, which coats the circuit board with a uniform conducting layer of copper, providing a base upon which additional copper can be deposited electrolytically. Circuit patterns can then be transferred onto the boards through photolithography or a stencil printing process. After photolithography, excess copper is removed using etchants such as sulfuric acid and hydrogen peroxide. The boards are then electroplated to build up the thickness and strength of the conducting layers and provide electrical conductivity between the layers of the PWB. The final steps in the preparation of PWBs involve soldering and testing.

Acid fumes and spent etchants are the primary acid-related pollutants. Source control techniques involve utilizing differential plating instead of the conventional electroless plating process, using thinner copper foil to clad the laminated board, and utilizing a pattern instead of panel plating.

In order to reduce the amount of copper that must be etched away, there is the possibility of differential plating. By controlling the concentrations of certain stabilizers in the electroless copper bath, copper should deposit three to five times faster on the through hole walls than on the copper cladded surface, reducing the amount of copper that must be etched away. Through the use of thinner copper foil in the cladding of laminated boards, the amount of copper that must be etched will be reduced. Pattern plating does not involve copper plating the entire board area, it merely requires copper electroplating only the holes and circuitry. Pattern plating is a more efficient plating method, leaving less copper behind in need of etching.

Inorganic Chemicals Industry

Acids and bases serve a variety of functions in the inorganic chemicals industry, SIC 281. For example, chlorine and caustic soda are co-products from the electrolysis of saturated solutions (salt water or brine) of sodium chloride, NaCl. In the electrolysis reaction that follows, small amounts of hydrogen gas are also produced:

$$2\ NaCl + 2\ H_2O \rightarrow 2\ NaOH + Cl_2 + H_2$$

DC electricity provides the necessary driving force for the reaction. There are three types of electrolysis processes used to produce chlorine, caustic soda, and hydrogen from brine: mercury cell, diaphragm cell, and membrane cell.

Brine solutions are treated with various chemicals for the removal of impurities prior to input to the electrolytic cells. In the case of mercury and membrane cell processes, brine is first acidified with HCl to remove dissolved chlorine. Sodium hydroxide and sodium carbonate are then added to precipitate calcium and magnesium ions as calcium carbonate and magnesium hydroxide. Barium carbonate is then added to remove sulfates as a barium sulfate precipitant. All impurities that have precipitated out of solution can then be removed by settling and filtration.

METAL FINISHING OPERATIONS

Metal finishing operations involve the preparation of metal parts for final use. These operations involve the alteration of the metal workpiece surface properties in order to increase corrosion and abrasion resistance, alter appearance, or in some way enhance the utility of the product. Metal finishing operations can include cleaning, degreasing, pickling (acidic removal of surface oxides), electroplating and electroless metal plating, etching, passivation, phosphating, chemical electropolishing, chemical milling, and anodizing conversion coating.

Proper cleaning of metal parts prior to further processing is key for effective coating adhesion. Typically, machined parts are cleaned with solvents. Paint, rust, and old plating is stripped from workpieces using caustics and abrasives. Workpieces in plating lines are cleaned several times using water, acids, caustics, and detergents.

Process Description

The design of an effective cleaning operation is dependent upon three factors: (1) the nature of the contamination; (2) the metal substrate; and (3) the degree of cleanliness required. Knowledge with regard to the nature of the contamination on the metal's surface and the effect the chosen cleaning agent and technique will have on the metal is important in order to design the proper cleaning system and sequence of baths or other operations. Also, some cleaners are more effective than others for removing a particular type of contaminant. For example, alkaline cleaners are often used to remove mineral oils, animal fats, and oils, while caustics are good paint stripping agents. Acid cleaners and abrasives are generally applied to remove oxidation scale and rust. Finally, to make matters difficult, parts that have been contaminated with several materials have to be cleaned by a sequence of operations.

An effective cleaning operation design depends on the characteristics of the metal substrate itself because the contaminant must be removed to the required degree without damaging the metal. Cleaners that are appropriate for one metal may not be for another because the reactivity of different materials with alkaline and acids varies. For instance, parts going to a cyanide-zinc plating bath do not usually require a preliminary high level of cleaning because cyanide-based plating solutions exhibit strong cleaning actions on their own. On the other hand, for a nickel plate to adhere to a metal surface, the surface must be extremely clean. Thus, thorough cleaning operations are required prior to nickel plating.

Solvents, used for removing oil-based contaminants in either cold cleaning, diphase cleaning, or vapor phase cleaning operations, are the most popular class of cleaners. The four categories of cold cleaning that usually employ unheated or slightly heated nonhalogenated solvents are: (1) wipe cleaning, (2) soak cleaning, (3) ultrasonic cleaning, and (4) steam gun stripping. Wipe cleaning and soak cleaning are self-explanatory. Ultrasonic cleaning is identical to soak cleaning, except that an ultrasonic unit is added to the tank, which provides turbulence and enhances cleaning action throughout the tank. In steam gun stripping, a stripper made up of nonhalogenated solvents is mixed with steam and ejected at high speed from a nozzle for the purpose of removing paint from metal objects.

In the electroplating process, an electric current is passed through a solution containing particular dissolved metal ions along with the metal substrate to be plated. The metal object is cathodic in nature and attracts metal ions from the solution. Ferrous and nonferrous metal objects are electroplated by a variety of metals. The electroplating bath solutions vary depending upon the metals to be utilized. Some common electroplating solutions are brass and bronze, composed of copper cyanide, zinc cyanide, sodium cyanide, sodium carbonate, ammonia, and Rochelle salt; copper cyanide, composed of copper cyanide, sodium cyanide, sodium carbonate, sodium hydroxide, and Rochelle salt; acid copper sulfate, composed of copper sulfate and sulfuric acid; chromium, composed of chromic acid, sulfuric acid, and possibly fluoride as a catalyst; and cadmium cyanide, consisting of cadmium cyanide, cadmium oxide, sodium cyanide, and sodium hydroxide.

A typical electroplating process involves a particular sequence of operations including cleaning, stripping of old plating or paint, the actual electroplating steps, and final rinsing. Generic process wastes include spent process solutions and filter sludges, from both plating and chemical conversions; quench oils and quench oil tank clean-up wastes, from case hardening; spent salt bath; waste water treatment sludge; vent scrubber wastes, from vent scrubbing; and ion exchange resin reagents, from demineralization of process waste.

Pollution Prevention Options

Cleaning Operations

An alternative in the utilization of solvents in workpiece preparation is to incorporate aqueous-based cleaning methods. Aqueous chemistries often consist of alkaline cleaning agents and require experimentation to determine effectiveness. In order to enhance the cleaning process, it may be reasonable to utilize pressure sprays or ultrasonics. The use of ultrasonic resonators results in an ultrasonic "scrubbing/wiping" action that improves cleansing efficiency.

Rinsewater Reduction

Countercurrent rinsing employs sequential rinse tanks in which the water flows in the opposite direction of the work flow. Fresh water is only added to the final rinse tank and is conveyed to the previous tank. This way, the material is first exposed to the dirtiest water for a preliminary cleaning, and is then progressively exposed to cleaner and cleaner water, improving the driving force for mass transfer. There is great potential in this approach for the minimization of water consumption.

Ion Exchange Technology

Short-bed ion exchange is a method that allows for recovery of metal ions due to dragout from the plating bath into the rinse baths. In cases where there are a series of rinse baths preceding the plating bath, the quality of water in the first rinse bath should be monitored with a conductivity probe. When the dragout of plating solution to the first rinse increases the conductivity above a set point, deionized water can be introduced to the third rinse. As the rinses counterflow forward, the rising level in the first rinse should then activate a level controller, which would activate a pump and transfer contaminated rinsewater to a feed tank for the recovery system.

The ion exchangers remove the dissolved metal from the rinsewater. The pH of the treated rinsewater can then be adjusted and the effluent discharged. The ion-exchange resins can be periodically regenerated automatically to yield a concentrated metal salt solution that can be recycled to the appropriate plating bath.

Ion-exchange metal recovery could be incorporated into a typical copper-plating process as follows. Rinsewater from the first rinsing tank following the plating bath, which contains a copper sulfate solution, can be transferred to a feed tank. A carbon column preceded by a cartridge filter can be used to remove additives and suspended solids. The effluent from these operations can then be fed through a column of fine mesh cation-exchange resin. In this column, copper cations in solution are exchanged for hydrogen cations on the resin. The cation exchangers can be regenerated with a dilute sulfuric acid solution pumped up through the column. After the initial volume, displaced rinsewater is diverted back to the feed tank and the sulfuric acid exchanges H^+s for copper cations. A concentrated copper sulfate solution would then be diverted to a holding tank for future usage in the plating bath, while excess acid and copper solution remaining in the column is washed out with water and combined with sulfuric acid regeneration in a holding tank.

FERMENTATION

The pharmaceutical industry utilizes various batch-type processes, one of which is fermentation. The major steps in a fermentation process are inoculum and seed preparation, fermentation, product recovery, and purification.

Process Description

In the preparation of sterile inoculum, microbial strains are matured into dense suspensions. Further propagation is allowed to take place in a seed tank, which operates like a full-scale fermenter. In the fermentation process itself, the fermenter is charged with material from the seed tank through a series of sterilized lines and valves. During the process, the vessel contents are usually aerated and agitated. The effective operation of a fermentation process relies heavily upon effective moni-toring of dissolved oxygen content, pH, and temperature. The solid residues left over after fermentation are usually removed by filtration. The filtrate is then processed to recover the desired product using solvent extraction, precipitation, and ion exchange technologies.

Pollution Prevention Options

Typical wastes generated in the fermentation process include spent aqueous fermentation medium, solid cell debris, and cleaning operations. Generally, the spent aqueous medium contains unconsumed raw materials such as corn steep liquor, fish meal, and molasses. Cell debris from filtration is usually in the form of spent filter cake, which includes the solid remains of the cells and residuals. Large quantities of wastewater are generated from equipment cleaning operations and fermenter vent gas scrubbing. Such waste waters usually have high BOD, COD, and TSS levels and a pH range of 4–8.

Areas of particular concern toward which pollution prevention and waste minimization efforts should be directed include solvent use and production efficiency. Recovery and recycling includes direct reuse of waste material, recovering used materials for separate use, and removing impurities from waste to obtain relatively pure substances. To improve solvent waste recyclability for cleaning and possibly product extraction procedures, solvents should be properly segregated. It should be noted, however, that there are regulatory parameters enforced by the U.S. Food and Drug Administration with regard to the recycling and reuse of solvents in pharmaceutical operations. As a general rule, solvents should be segregated so as to keep chlorinated away from nonchlorinated solvent wastes, aliphatic from aromatic solvent wastes, chlorofluorocarbons from methylene chloride, and water wastes from flammables. Processes available for solvent recovery include distillation, evaporation, liquid–liquid extraction, sedimentation, decantation, centrifugation, and filtration. Waste generation due to the production of off-spec batches can be reduced through the use of automated processing systems and through minimizing the use off-spec or outdated products.

WASTEWATER TREATMENT

Wastes are generated by practically all industrial operations. Since disposal costs of solid residues correlate directly with mass, volume, and toxicity; treatment before disposal is desirable so as to reduce such characteristics. Two possible options that can be implemented in order to produce less sludge in waste treatment operations are treatment chemistry modification and sludge dewatering.

Pollution Prevention and Wastewater Sludge Reduction

Exemplary of reducing sludge via a chemical modification approach is that developed through research sponsored by the U.S. Air Force. The conventional treatment of a mixed metal waste containing hexavalent chromium initially calls for the reduction of pH to below 3 through the addition of mineral acid. Then a reducing agent is added to convert hexavalent chromium to the trivalent state. Lime is then added to raise the pH to approximately 9.5 so as to precipitate heavy metals as hydroxides. This process results in the production of large quantities of sludge due to the fact that calcium carbonate and calcium sulfate precipitate out as well. In the new approach, a combination of ferrous sulfate and sodium sulfide is used as a reducing agent and sodium hydroxide is utilized to precipitate heavy metals. The efficiency of precipitation is now greatly enhanced in that iron hydroxide precipitate removes other heavy metals as well, and at a neutral to slightly alkaline pH. This reduces the need for lime and minimizes the production of calcium sulfate and carbonate solids.

Sludge dewatering can be accomplished with three types of mechanical devices: vacuum filters, belt presses, and plate-and-frame filter presses. Vacuum filters make use of diatomaceous earth to improve dewatering. It is noteworthy that during the solids capturing process, substantial quantities of diatomaceous earth are often scraped off, adding to the overall quantity of solids requiring disposal. With the plate-and-frame filter press, which generally operates in a batch mode, the solids are retained by a synthetic cloth filter media. Belt presses differ in that they are capable of operating continuously.

BLEACHING PROCESS

Bleaching is the addition of chemicals in a staged process of reactions and washing, to increase the whiteness and brightness of a material. Bleaching operations are common to the pulp and paper, textiles, and photoprocessing industries. The pulp and paper industry is most significant for bleaching pollution concerns.

Process Descriptions

Pulp and Paper Industry

The bleaching process in the pulp and paper industry (SIC Code 26) is primarily used to whiten and brighten pulps for paper manufacture and may produce wastewaters containing chlorinated compounds such as dioxins. Chlorine reacts with organics to form trihalomethanes in receiving waters. Overall, the pulp- and paper-making industry is water intensive; it is the largest industrial user of process water in the United States. Pulp and paper mills usually operate wastewater treatment plants to remove biological oxygen demand (BOD), total suspended solids (TSS), and other pollutants before discharging wastewaters to receiving waterways. Mills may also discharge into public-owned treatment works (POTW).

There are four major chemical pulping techniques:

1. Kraft or sulfate
2. Sulfite
3. Semichemical
4. Soda

In the pulp and paper industry the kraft pulp process involves highly bleached and purified kraft process wood pulp suitable for conversion into products such as rayon, viscose, acetate, and cellophane.

The kraft or sulfate process produces over 80% of the chemical pulp products. Bleached or unbleached wood pulp from the kraft process is usually converted into paper board, coarse paper, tissue paper, and fine paper products for writing.

Paper-grade sulfite-processed wood pulp, with or without bleaching, is used for products such as tissue paper, fine paper, and newsprint. As in the kraft technique, dissolving sulfite uses highly bleached and purified wood pulp suitable for conversion into products such as rayon, viscose, acetate, and cellophane.

Semichemical pulp is produced by chemical pressure and mechanical forces with or without bleaching. This pulp is used for corrugating medium for cardboard, paper, and paperboard products. After fibers are separated and impurities are removed, the pulp may be bleached to improve brightness and processed to a suitable form for paper making. Most of the pollution released with paper mills comes from the pulping and bleaching processes, where the majority of chemicals are added.

Unbleached pulp is mostly used in boxboard, bags, and linerboard. Bleached pulp is needed for paper where whiteness and brightness are preferred, as in printing and business paper. Almost any pulp can be bleached, but the fiber type affects the degree of bleaching possible. Lignin content in pulp is the determinant for bleaching potential. A high lignin content has a greater bleaching potential and yield.

Chemical pulps are bleached in bleach plants where the pulp is processed through three to five stages of chemical and washing cycles. The number of cycles depends on the whiteness desired. Bleaching stages alternate between acid and alkaline conditions. Alkaline extraction stages dissolve the lignin and the acid produces chemical reactions with the lignin. The washing stage removes solutions and reaction products. Commonly 4% to 8% of pulp is lost due to bleaching agent reacting with wood cellulose fibers.

The most common chemicals used in the bleaching process for paper are:

- Chlorine dioxide
- Elemental chloride
- Hydrogen peroxide
- Hypochlorite
- Sodium hydroxide
- Sulfur dioxide
- Sulfuric acid

Water pollutants from the pulp and paper bleaching process include BOD, color, and chlorinated organic compounds. The potential environmental hazards from washwater sludges are associated with trace constituents like chlorinated organic compounds that form in the effluent. Pollution outputs may differ based on mill process and material inputs (e.g. wood chips, resin content).

Textiles

Chlorine and oxygen bleach baths are also used in the manufacturing of textiles (SIC Codes 22 and 23). Products that may use bleaching in the textile industry are wool, cotton, and synthetic fibers. Following desizing and scouring, bleaching of cloth with chlorine (Cl_2 (aq)), hypochlorite (OCl^- ion), or peroxide (H_2O_2) may be performed to whiten the fabric to a preferred degree of color. Synthetic fibers are man-made, such as acetate, acrylic, nylon, polyester, and rayon. Bleaching and preparation processes contribute very little to the overall plant BOD emissions. Synthetic fibers and blends require little or no bleaching. Wetting agents and other additives in the bleaching formulations are a minor contributor to the BOD problem of textile mills.

Cloth is first rinsed with water and then rinsed in an antichlor bath containing sodium bisulfite and weak solutions of sulfuric or hydrochloric acid. The washing is followed by another rinse with washwater and the cloth is passed through a hypochlorite solution, squeezed lightly, and stored in bins for about 24 hrs at constant temperature between 55 and 65°F for drying.

Wool is bleached with sulfur dioxide or hydrogen peroxide to whiten the natural yellow tint and remove debris. Synthetic fibers may be bleached only for color quality since they are free of impurities. Toxic substances from bleaching include chlorine and hydrogen peroxide.

Photographic

The photographic processing operations use permanganate (MnO_4^-ion), ferricyanide ($(Fe(CN)_6)^{-3}$ ion), and dichromate (CrO_7^{-2} ion) bleaches. Ferricyanide is a concern due to potential for decomposition into free cyanide in sunlight. The photoprocessing industry is based on facilities that develop and finish photographic film. Most consumer films are based on silver as the photoactive chemical. Bleaches are used to remove the black silver to produce the film image. Spent bleach contains silver and is regarded as hazardous waste. In black-and-white film processing, spent bleach is hazardous because of its chrome content.

The processing of film requires the use of many chemicals resulting in a broad chemical wastestream. Color processing requires a bleach bath to produce the color image visible by removing the black metallic silver image, converting the metallic silver back to a silver halide. The silver is then dissolved and removed in a fixer bath. Some processes combine the bleaching and fixing processes in a single solution (bleach-fix or "blix").

Black and white reversal film processing requires two development steps with an intermediate step. Bleach solution for black-and-white processing contains sodium dichromate and spent solutions are considered hazardous from chrome content. Waste generated by photoprocessors are primarily aqueous effluents from process bath wastes, color developer wastes, and bleach-fix wastes. Ferricyanide bleach results in ferricyanide constituents that are toxic to the environment.

Dichromate bleach results in organic chemical and chromium compounds with oxygen demand and toxic metal concerns. Ferricyanide bleach is used primarily in cinemagraphic films. Most processors recover up to 99% of the ferricyanide for reuse. Ferrocyanide will convert into free cyanide in sunlight.

Pollution Prevention Options

Pulp and Paper Industry

Major pollution prevention is in the area of reduction of toxic and chlorinated compounds. Most conventional, end-of-pipe treatment technologies are not effective in destroying all chlorinated compounds. Source reduction and material substitution techniques are the main area of concentration in preventing chlorinated releases. Defoamers, bleaching chemicals, or wood chip substitution may reduce the use and releases of chlorinated compounds.

Extended delignification reduces the lignin content before the pulp moves to the bleaching station. By reducing the lignin content, the amount of bleaching agents needed will be reduced. The extended process involves: increasing cooking time, adding cooking chemicals at several points while cooking, regulating cooking temperatures, and controlling the concentration of hydrogen sulfide ions and dissolved lignin. Color, BOD, COD, and chlorinated compounds generated during the bleaching process are reduced.

Oxygen delignification can also reduce the lignin content in the pulp, which involves the addition of an oxygen reactor between the kraft pulping stage and the bleach plant. The amount of bleaching chemicals required for whiteness is directly connected to the amount of lignin contained in the pulp. The pulp is fluffed, steam heated, and injected with gaseous oxygen. After oxygenation, the pulp is washed to remove lignin prior to bleaching. The process can be used with other process modifications to completely eliminate the need for chlorine-based bleaching agents.

Ozone delignification (ozone bleaching) is being used to eliminate the need for chlorine in the bleaching process of paper pulp. This technique must be conducted at low pH (1.0 to 2.0), requiring sulfuric acid, and at constant temperatures of 20°C. Organic materials must be washed out of the brownstock pulp and ozone reactive metals removed. Overall emissions are lower than conventional processes because effluents from each stage can be recycled. Ozone generation must be generated onsite, resulting in increased energy costs. The increased energy cost should be offset by the elimination of chlorine production and chlorine dioxide use.

The addition of anthraquinone catalysts (chemical from coal tar) to the pulping liquor may speed up the kraft pulping reaction and increase yield by protecting cellulose fibers from degradation. This lowers the amount of lignin in the prechlorination pulp, reducing bleaching chemicals needed.

Improving washing can reduce the amount of bleaching chemicals and can result in reductions in chlorinated compounds as well as conventional pollutants. New washing systems with improved solids removal and energy efficiency are beginning to replace the conventional rotary vacuum washers. Most new mills are designed with countercurrent washing, systems while others are modernizing older wash systems.

The substitution of chlorine dioxide for elemental chlorine as a bleaching agent is being used for its improvement on pulp and effluent quality. The use of chlorine dioxide in place of chlorine increases oxidative reaction proportions, reducing the formation of residual chlorinated organic pollutants. By substituting chlorine dioxide, chlorinated compounds are reduced by 80%. A substitution of 50% to 70% in the first bleaching stage is now a common practice. The use of chlorine dioxide is more expensive than that of chlorine powders.

Improving wood chipping and screening can result in reductions of bleaching chemicals used. Incorporating equipment that will separate chips by size and thickness will reduce the quantity of chemicals needed for the removal of lignin.

Oxygen-reinforced and peroxide-extraction processes used together or separately can reduce the amount of elemental chlorine and chlorine dioxide needed for bleaching while improving pulp quality.

Improving chemical controls and mixing can minimize excess chlorine bleaching chemicals within the reactors. Chemical application control and monitoring systems can decrease formation of chlorinated organic compounds. Although many of the pollution prevention techniques are aimed at reducing toxic release, the process change can lead to reductions in Biological Oxygen Demand (BOD), Total Suspended Solids (TSS) and Chemical Oxygen Demand (COD). Also, reductions in water use, air emissions, and sludge volumes are contributed from process changes.

Textiles

Reconstituting bleach baths for further use before discharging will significantly reduce the volume and concentrations of the plant effluent. Bleach baths, particularly with synthetic fibers, may be used many times to reduce pollution while saving raw material costs.

Chemical conservation reduces unnecessary chemical use, which may affect effluent quality and processing costs. The reuse of bleach washwater for scouring or desizing operations may be performed if size recovery is not a priority.

A good housekeeping and maintenance plan should be established for pollution prevention. The reductions in leaks, spills, and equipment failure will increase productivity while reducing unnecessary wastes.

Photographic

Material substitution can reduce toxic waste streams. For example, ferricyanide bleach can be replaced with ferric EDTA (ethylenediaminetetraacetic acid) complex, resulting in a less toxic waste.

Squeegees may be used in all manual and some automated processing systems to wipe off excess liquids from the film and paper, reducing chemical carryover between bath solutions. Squeegee use results in longer bath life and low chemical contamination.

Silver recovery from fixer and bleach-fix solutions can result in economic gain in the silver trades. As much as 80% of silver processed for black-and-white film and nearly 100% of the silver in color film processing end up in the fixer or bleach-fix solutions. The most common recovery techniques for silver are electrolytic recovery, ion exchange, chemical precipitation, and reverse osmosis.

Ferricyanide bleach reduces to ferrocyanide during the bleaching process. The spent ferrocyanide can be regenerated and reused in photo processing. Recovery methods may be ozone regeneration, persulfate regeneration, or electrolytic regeneration. To reduce photographic chemicals and silver, use silver-free films like vesicular, diazo, or electrostatic types. Water-developed litho plates may also be used.

5

The Internet: A Great Starting Point for Pollution Prevention

Don't overlook the Internet as a source of pollution prevention information. The Internet is a global network of computer networks that provides the user of a connected computer with access to information sources all over the world. In providing quick access to all sorts of documents, the Internet proves itself an excellent starting point in the search for environmental information. As you progress through this chapter, you may come across some unfamiliar technical terminology. Fret not; there is a list of terms, with explanations, at the end of the chapter.

The Internet offers users of personal computers with connections to the "network of networks" access to an ever-increasing quantity of information. Point-to-Point Protocol (PPP) is an example of a set of rules that allows for such required connections and communications to networks over telephone links. Usually software is required so that a PPP link can be made between a personal computer and the computer system of an Internet Service Provider (ISP). ISPs are companies or universities that provide user accounts within their network, which is connected to the Internet. It should be made clear that PPP is only one of several different ways to connect to your ISP. Further research about the best connection option for a particular location will be left up to the reader. Again, PPP allows for links between personal computers and ISPs to be established. After a link is established, standard Internet protocols such as Transmission Control Protocol/Internet Protocol (TCP/IP) and File Transfer Protocol (FTP) become involved. TCP/IP governs data transmission, while FTP governs the way files are transferred from one computer to another, over phone lines, and across the Internet.

Accessing the Internet is becoming an easier, more efficient process every year. Recently, modem speed has been increasing on an almost yearly basis. Cable companies are working on delivering Internet access via their coaxial lines, and telephone companies are improving ISDNs (Integrated Services Digital Networks) and ADSLs (Asymmetric Digital Subscriber Lines). What this means for the user is that internet access is not only becoming easier, but faster and more reliable as well.

For now let's take a look at some of the basics for getting connected to the Internet. First of all, you will need a modem. Modems are like telephones for the computer, dialing numbers and converting data into a transmittable format that other modems can understand. Next you'll either need terminal emulation software for simple communications or Internet application software, inclusive of TCP/IP stack, or both. Internet application software, generally in the form of a web browser, supports a graphical user interface that provides the user with a comfortable medium through which to view information available on the Internet. Finally, you'll need to maintain an Internet account with a local provider's network or through a subscription to an Internet service provider (such as America Online). Generally, your Internet Service Provider will furnish you with the required software and will assist you in getting started.

INTERNET TOOLS

Today's Internet provides access to a number of basic services: electronic mail, interactive conferences, information resources, network news, and the ability to transfer files. There are also a number of tools available to guide the user through the library-like system of the Internet.

E-mail

Electronic mail, or e-mail, is a communication tool that enables a user to send and receive messages or "mail." In order to send a message, all you need is the receiver's e-mail address, which consists of a username and a host/domain name separated by the @ symbol.

USENET

Also called NetNews or Newsgroups, USENET is a giant bulletin board system, composed of newsgroups (discussion forums) organized by topic. To read newsgroups, you will need a newsreader, which can be a stand-alone program or one that is integrated into modern Web browsers. Anyone with access to newsgroups through an ISP can join in at any time, but its best to "lurk" (read others' posts) before joining in so you can avoid repeating frequently asked questions (FAQ) and annoying the regulars. If you are unfamiliar with

USENET, two newsgroups that provide more information are news.announce.newusers and news.newusers.questions.

World Wide Web

The World Wide Web (WWW) is an information system based on "hypertext." It provides a means of moving from document to document within a network of information. It should not be confused with the Internet itself; it is merely a part of the Internet. The WWW has become a major part of the Internet because of its hypertext/hypermedia capabilities. Through the use of software called a "browser," the "web surfer" can jump from web page to web page by selecting coded "links" with a mouse or by typing in a web address. As Web browsers have become more sophisticated, they have gained the ability to access virtually all parts of the Internet, not just those that are part of the World Wide Web. For this reason, the terms Internet and World Wide Web are now often used interchangeably.

Telnet

Telnet (short for telephone network) lets the user log into another Internet-connected computer. It gives the user the ability to obtain information from remote databases all over the world. When using Telnet, your own computer is acting like a terminal, sending commands to and receiving information from the remote computer that executes the commands. Telnet is a convenient way to tap into Bulletin Board Systems (BBSs) even if you do not subscribe. Generally, using Telnet is simple; at the prompt, just type "telnet" followed by the site you wish to visit.

FTP

FTP (short for File Transfer Protocol) is a method for transferring files from one computer to another through the Internet. There are many tools available that will allow you to FTP files. In many cases, the easiest way to perform file transfers is to use a Web browser, which has built-in FTP capabilities. Although Web browsers are probably sufficient for most transfers, there are also stand-alone FTP programs that can offer some extra versatility.

Archie

Archie is a tool for finding specific files on FTP servers. To use Archie, you need to know the name of the computer file, or part of the name. Archie searches its directory of FTP sites for individual, publicly available files and documents. These files are often software programs of related pieces of documentation. Archie servers can be accessed using a Web browser.

Gopher

Gopher is a menu-driven system that can be utilized to access documents, information, and services available on the Internet. Gopher was the Internet's main interface before World Wide Web browsers were developed. Today, you can connect to a Gopher site using a text-based browser or a Web browser such as Netscape Navigator®. With a browser, one has to simply enter the Universal Resource Locator (URL) address in the browser's location box. Upon arrival at the gopher site, merely scan the menu using the scroll bar at the right of the window.

Veronica

Veronica (an acronym for Very Easy Rodent-Oriented Netwide Index to Computer Archives) permits the user to search "Gopherspace" by keyword, and is useful if you wish to learn more about a subject but don't know where in Gopherspace the information resides. There is usually a selection for Veronica available from the menu at the particular Gopher site you are visiting. Veronica performs Boolean Searches, allowing you to use connectors like AND, OR, and NOT to combine words or strings of words. Unlike search engines (see below), Veronica is limited in that it searches only Gopher menus and submenus rather than entire files, and indexes only those files that are located on Gopher servers.

WAIS

WAIS (Wide-Area Information Servers) is another search tool for finding information on the Internet. WAIS is available at many World Wide Web sites and can be conveniently utilized by means of the Web browser. You simply enter a keyword and the program automatically searches connected databases and returns a list of those databases that contain the search term. Although WAIS searches entire files, the searches are limited to those files that are included in WAIS databases.

IRC

Using IRC (Internet Relay Chat), you can participate in real-time discussions over the Internet. The discussion groups, called channels or chatrooms, come and go as they are created by the participants, and a channel remains active as long as there is one participant using it. For more information, check out the newsgroup: alt.irc.

RESEARCHING AND EVALUATING POLLUTION PREVENTION OPPORTUNITIES

The Internet, especially such tools as the World Wide Web and e-mail list-server, is a great place to start when researching and evaluating pollution prevention, waste minimization, and pollution control. This section contains references to a number of Web sites. For your convenience, there is also a list of these sites along with the corresponding URLs.

When looking for ideas and professional insight into your particular situation, an e-mail list-serve could be a great place to start. An e-mail list-serve is a system that allows you to send e-mail to a large group of individuals who subscribe. You simply direct your e-mail to the list-serve address and a copy of that message winds up in each subscriber's mail box. In addition, each subscriber's response to that message will wind up in the other subscribers' mailboxes, inclusive of the original sender's. It should be noted that access to some list-serves, such as P2-Tech, may be restricted, and potential participants for these are usually screened when they "subscribe." However, keep your eyes and ears open for new list-serves. You may even wish to start one among fellow members of a trade organization or other professional affiliation.

Next you may want to peruse the World Wide Web for information on environmental regulations, pollution prevention and control technologies, and hazardous materials. In refining your search, you may wish to start off using a search engine such as Alta Vista. Search engines are programs that allow you to perform keyword searches for information available on the Web. Search engines are further described later in this chapter.

Finally you may wish to access particular sites that are known to have specific information related to your needs. For example, federal regulatory information is available at the USEPA's website. In addition, state and local updates are available from a multitude of websites maintained by law offices. Also, many organizations that do not maintain Web sites do support a Bulletin Board System (BBS). BBSs utilize a DOS-based, menu-driven system and allow you to view and download files. To access a BBS you need a modem and an all-purpose telecommunications software such as Procomm Plus®. It doesn't hurt to call an organization of interest to ask if they support a BBS.

Further, many individual companies, state technical assistance programs for pollution prevention, universities, and trade associations support Web sites providing access to pollution prevention technology resources. In the list that follows, such germane Web sites are presented along with their respective URLs and a brief description. They are divided into the following categories:

- Government and regulatory
- Nonprofit organizations, research and information centers, and technical assistance programs
- Business and industry
- Professional organizations
- Search engines

Enjoy!

WWW Resources

It should be noted that WWW sites are continuously evolving and that URLs frequently change. The following sites were verified as of February 3, 1997. If you come across an incorrect URL, don't give up—try one of the search engines.

Government and Regulatory Web Sites

Enviroene

http://es.inel.gov/index.html

Supported by the USEPA, Enviroene is a very comprehensive site that attempts to provide a single repository for pollution prevention, compliance assurance, and enforcement information and databases. At the Enviroene site, you can find P2 case studies, technologies, points of contact, environmental statutes, executive orders, regulations, and compliance and enforcement policies and guidelines. There is also a database containing solvent alternative information.

Envirotext

http://tamora.cs.umass.edu/info/envirotext/
An on-line searchable library that provides easy access to environmental regulations, Envirotext is a self-proclaimed "one-stop-shop" for all your environmental regulatory needs.

SAGE: Solvent Alternatives GuidE

http://clean.rti.org/
An on-line comprehensive guide maintained by the Research Triangle Institute, SAGE is designed to provide pollution prevention information on solvent and process alternatives for parts cleaning and degreasing. The site also includes links to a number of Web sites dedicated to environmental and pollution prevention information, and a link to CAGE, the Coatings Alternative Guide.

United Nations Environmental Programme's (UNEP) World Wide Web Site

http://www.unep.no/
Supported by the UNEP, this site offers information on environmentally sound technologies.

U.S. Environmental Protection Agency

http://www.epa.gov/
This site provides public access to EPA information, including announcements, policy, regulations, environmental test methods and guidelines, research, technology, and databases.

Nonprofit Organizations, Research and Information Centers, and Technical Assistance Programs

Battelle Seattle Research Center

http://www.seattle.battelle.org/
This site is supported by the Technology Management and Social Sciences Division of the Battelle Seattle Research Center. The organization is home to many technical and performance management professionals dedicated to providing practical solutions to organizational and technical challenges. Of particular interest at this site is the list of P2 and environmental links, a page entitled "Finding and Using Pollution Prevention Information on the Internet." To get to this page, start at the home page and click on the link for "text outline of our site's contents." Upon arriving at the new page, click on "Pollution Prevention" under "Environment and Society." You will then arrive at a page that includes another comprehensive listing of links categorized as follows: Pollution Prevention Technology Resources, Design for Environment (DfE) information, Recycling Information, State Pollution Prevention Programs, Regulations and Policy Information, and Trade and Professional Associations.

Center for Technology Transfer and Pollution Prevention: CT2P2

Los Alamos National Laboratories, Materials Substitution Resource

http://ingis.acn.purdue.edu:9999/cttpp/cttpp.html
Located at Purdue University, the center provides environmental professionals with the tools necessary to transfer technical information about the environment and P2 worldwide. CT2P2 has developed a series of over

thirty computer programs on various aspects of the environment and pollution prevention.

Environmental Stewardship

Los Alamos National Laboratories, Materials Substitution Resource

http://perseus.lanl.gov/
This site provides access to pollution prevention publications and a comprehensive list of environmental links. When you arrive at the home page, simply click on "P2 Information by Topic on the Internet" and gain access to links to Material Substance Information, MSDS/Chemical Information, Recycling Information, Ozone Depletion/Global Warming Information, and State Environmental Programs.

Great Lakes Information Network (GLIN)

http://www.great-lakes.net/
GLIN is a collaborative project of agencies and organizations in the binational Great Lakes region to link data, information, and people via the Internet. This organization also supports the aforementioned list-serve, P2-Tech (see below). GLIN should be your first stop for environmental information relating to the Great Lakes region, but has a wealth of information that is applicable to other regions as well.

Minnesota Technical Assistance Program (MnTAP)

http://es.inel.gov/techinfo/facts/mpca/mpca.html
This site facilitates access to a large number of fact sheets that are great pollution prevention research materials.

New Jersey Technical Assistance Program

http://www.njit.edu/njtap
This site facilitates access to great P2 reference material. It also supports a search engine, EnviroDaemon, which

is designed to streamline the P2 data searching process. Check it out!

North Carolina Division of Pollution Prevention and Environmental Assistance (DPPEA)

http://owr.ehnr.state.nc.us/
A great web site for information on pollution prevention, technical assistance, and recycling.

Ohio Office of Pollution prevention

http://www.epa.ohio.gov/opp/oppmain.html
This site contains a wealth of P2 information and resources, as well as an on-line pollution prevention assistance program where P2 questions are answered via e-mail.

The P2-Tech Archives

http://gopher.great-lakes.net:2200/1s/mailarc/p2tech
Supported by GLIN, this site provides access to archives from the P2-Tech list-serve and a variety of other references.

The Right-to-Know Network

http://rtk.net/
This site provides free on-line access to over 100 gigabytes of quantitative databases and numerous text files and conferences on the environment, housing, and sustainable development. The databases are derived from the USEPA, the U.S. Department of Housing and Urban Development, the U.S. Federal Election Commission, and the U.S. Bureau of the Census. It is jointly maintained by two nonprofit organizations, OMB Watch and The Unison Institute.

The Tellus Institute

http://www.tellus.com/
Supported by GLIN, this site provides access to archives from the P2-Tech list-serve and a variety of other references. The Tellus web site contains a great deal of information on pollution prevention and environmental accounting, as well as other environmental topics.

Business and Industry Web Sites

Dow Chemical Company

http://www.dow.com/
This site provides company-specific information, including products and services, environment health and safety, and financial information.

Environ/Net

http://www.teleport.com:80/~environ/
Environ/Net offers an electronic index of companies and organizations engaged in providing leading environmental technologies.

Monsanto

http://www.monsanto.com/
The Monsanto site includes information on environmentally related issues, careers, and specialty products.

MSDS Information

http://www.hazard.com
Vermont SIRI (Safety Information Resources on the Internet) claims to have the world's largest freely accessible MSDS (Material Safety Data Sheets) archive. The Web site includes links to many other MSDS archives, as well as other useful information.

SIC Codes

http://www.osha.gov/oshstats/sicser.html
One of several Standard Industrial Classification (SIC) Code cross references, this page is maintained by the U.S. Department of Labor Occupational Safety and Health Administration.

Switchboard

http://www.switchboard.com/
Switchboard contains no P2-related information, but is a great place to go for business telephone numbers all over the United States.

Thomas Register of American Manufacturers

http://www.thomasregister.com:8000/index.html
An on-line database for products and service suppliers, this site is easier and faster to use than the unabridged book version.

PROFESSIONAL ORGANIZATIONS

Air and Waste Management Association

http://www.awma.org/
This site is maintained by the Air and Waste Management Association, a nonprofit technical, scientific, and educational organization.

Environmental Professional's Homepage

http://www.clay.net/
This site has a very comprehensive set of links designed specifically for environmental consultants and remediation professionals. It presently highlights those difficult-to-find, critical links associated with regulations, government agencies, and health and safety issues.

National Center for Manufacturing Sciences

http://www.ncms.org/
This site—supported by NCMS, a consortium of U.S. and Canadian corporations promoting next-generation technologies—also supports a great Web page for the National Metal Finishing Resource Center.

SEARCH ENGINES

Imagine yourself in a library with millions of volumes on thousands of disparate topics. But these books are not organized by subject, as they would be in an ordinary library. They are scattered, seemingly at random, and with titles that may or may not give some hint as to their actual content. The Internet is like that library, but worse. It can give hours of enjoyment as you jump from site to site, exploring whatever you may find; however, finding the specific information you need can sometimes be quite frustrating. Luckily, there are search engines to guide you in your search for information.

Search engines are basically programs, available at particular web sites, which continuously scan the Web, monitoring and cataloging information. Type in your keywords (most search engines allow Boolean searches), and receive a list of Web sites that support your search criteria. A few of the many search engines are listed below.

AltaVista

http://www.altavista.digital.com/
AltaVista provides access to the largest Web index: 31 million pages found on 476,000 servers, and four million articles from 14,000 USENET news groups.

EnviroDaemon

http://www.njit.edu/njtap/
Accessible from the home page of the New Jersey Technical Assistance Program for Industrial Pollution Prevention, EnviroDaemon searches over 25,000 files on approximately 150 environmental and P2-related sites. Because it limits the search to preselected environmental sites, you should receive fewer, but more relevant, results than from the more standard search engines.

Infoseek

http://www.infoseek.com/
Infoseek allows you to type a question in plain English. It will recognize names and phrases to help you find exactly what you are looking for. You may get thousands of results, but Infoseek will display the best matches first.

P2 Gems

http://www.turi.org/P2GEMS
Developed by the Massachusetts Toxic Use Reduction Institute, P2 Gems can aid facility planners, managers, and engineers who are looking for technical P2 information. Over 125 sites with information relating to pollution prevention and toxics use reduction have been catalogued. You can search by keyword or by category. This is a good source for P2 research documents and summaries, as well as the names of experts and resources.

Yahoo!

http://www.yahoo.com
Yahoo! is one of the oldest Web search engines, and still one of the best. You can search the entire Web, or search one of the many categories and subcategories the people at Yahoo! have provided.

TERMINOLOGY

FAQ Short for frequently asked question lists, these are Web pages filled with useful tips and techniques.

File Transfer Protocol (FTP) FTP defines the rules governing the way files are transferred across the Internet, over phone lines, and from one computer to another. FTP also is often the name of a file-transfer program, through which Web browsers can download files from Web sites. There are two types of FTPs, a regular (private) FTP and an anonymous FTP. The private FTP is a restricted server that requires a username and a password to enter, while the anonymous FTP is open to anyone. For an anonymous FTP the username is usually "anonymous" and the password is either your e-mail address or "guest."

Gopher Before the World Wide Web, Gopher was a framework used for connecting to Internet resources that didn't require users to be network gurus. It uses character-based menus to provide access to data files such as programs, text, and images. Because it requires no special graphics, Gopher is still encountered frequently when browsing the Web.

Home Page A home page is a starting point for a service or an entity on the World Wide Web that usually features a combination of text, graphics, sound, and video that contain hyperlinks to other Web pages.

Hyperlinks Hyperlinks are objects (text or graphics) on a page that transport you to other areas, either within a home page or to another location, when you click on them.

Hypertext Markup Language (HTML) HTML is a set of the codes used to create Web documents that determine how information is shown on the screen.

Hypertext Transfer Protocol (HTTP) HTTP defines the rules for requesting and sending information over the World Wide Web. Your web browser sends out requests for information, and Web servers you connect to must respond with the data you requested.

Internet Relay Chat (IRC) IRC is an on-line forum that allows users to participate in real-time discussions on topics of interest.

Point-to-Point Protocol (PPP) PPP describes the rules for connecting computers to networks over telephone links. When you set up an Internet account, your computer may have to run special software to create a PPP link with your Internet Service Provider's computer. PPP is one of several protocols for connecting with your ISP.

Serial Line Internet Protocol (SLIP) SLIP is an old method of connecting computers to the Internet over telephone lines. While many people still use SLIP, PPP is preferable.

Server A server is a computer offering network services, such as storing contents of Web pages for browsing by anyone connected to the Internet.

Transmission Control Protocol/Internet Protocol (TCP/IP) The most basic underlying rules governing how data is transmitted around the Internet are defined in TCP/IP. This protocol is required for communicating between systems connected to the Internet.

USENET Newsgroups Similar to discussion groups on bulletin board systems (BBSs), there are newsgroups for every topic.

Web browser A Web browser is a software program for accessing text, graphics, and other data stored on the World Wide Web.

World Wide Web (WWW) World Wide Web is the name given to the interconnected system of computers that provides information using the HTTP and HTML protocols. Because Web servers link to other Web servers, the result can be visualized as a "web" of connections between all the servers around the world.

Appendix: Trade and Professional Associations

The Aluminum Association (AA)
900 19th Street, NW
Washington, DC 20006
Tel.: (202) 862-5100
Fax: (202) 862-5164
WWW: http://www.aluminum.org

Aluminum Recycling Association (ARA)
1000 16th Street, NW, Suite 400
Washington, DC 20036
Tel.: (202) 785-0951

American Automobile Manufacturers Association (AAMA)
1401 H Street, NW, Suite 900
Washington, DC 20005
Tel.: (202) 326-5500
Fax: (202) 326-5567
WWW: http://www.aama.com

American Ceramic Society
735 Ceramic Place
Westerville, OH 43081
Tel.: (614) 794-5817
Fax: (614) 899-6109
WWW: http://www.acers.org

American Chemical Society
1155 16th Street, NW
Washington, DC 20036
Tel.: (202) 872-8724
Fax: (202) 872-6206
WWW: http://www.acs.org

American Concrete Institute (ACI)
38800 Country Club Drive
Farmington Hills, MI 48331
Tel.: (810) 848-3700
Fax: (810) 848-3701
WWW: http://aci-int.org/

American Electronics Association (AEA)
5201 Great American Parkway, Suite 520
Santa Clara, CA 95054
Tel.: (408) 987-4200
Fax: (408) 970-8565
WWW: http://www.aeanet.org

American Institute of Chemical Engineers
1707 L Street, NW, Suite 333
Washington, DC 20036
Tel.: (800) 242-4363
Fax: (202) 833-3014
E-mail: dc@aiche.org
WWW: http://www.aiche.org

American Petroleum Institute
1220 L Street, NW
Washington, DC 20005
Tel.: (202) 682-8000
Fax: (202) 682-8031
E-mail: api@api.org
WWW: http://www.api.org

American Society for Surface Mining and Reclamation (ASSMR)
21 Grandview Drive
Princeton, WV 24740
Tel.: (304) 425-8332

American Trucking Associations
2200 Mill Road
Alexandria, VA 22314
Tel.: (703) 838-1844
Fax: (703) 838-1992
WWW: http//www.trucking.org

American Waterways Operators
1600 Wilson Boulevard
Suite 1000
Arlington, VA 22209
Tel.: (703) 841-9300
Fax: (703) 841-0389

Association of Hazardous Waste Materials Transporters
2200 Mill Road
Alexandria, VA 22314
Tel.: (703) 838-1703
Fax: (703) 519-1866
E-mail: 103424.1150@compuserve.com
WWW: http://www.tsdcentral.com/awhmt

Association of Iron and Steel Engineers
3 Gateway Center, Suite 2350
Pittsburgh, PA 15222
Tel.: (412) 281-6323
Fax: (412) 281-4657
WWW: http://www.aise.org/

Association of Rotational Molders (ARM)
435 North Michigan Ave, Suite 1717
Chicago, IL 60611-4067
Tel.: (708) 571-0611
Fax: (630) 571-0616
E-mail: 104360.240@compuserve.com
WWW: http://www.village.com/arm/

Business and Institutional Furniture Manufacturers Association
2680 Horizon Drive, SE, Suite A-1
Grand Rapids, MI 49546
Tel.: (616) 285-3963
Fax: (616) 285-3765
E-mail: email@bifma.com
WWW: http://www.bifma.com

Chemical Manufacturers' Association
2501 M Street, NW
Washington, DC 20037
Tel.: (703) 741-5000
Fax: (703) 741-6000
WWW: http://www.cmahq.com/

Copper Development Association (CDA)
2 Greenwich Office Park
Box 1840
Greenwich, CT 06836

Tel.: (212) 251-7200
Fax: (212) 251-7234
E-mail: The-Copper-Page@copper.org
WWW: http://www.copper.org

Environmental Conservation Board of the Graphic Communications Industries
1899 Preston White Drive
Reston, VA 22091-4367
Tel.: (703) 648-3218
Fax: (703) 648-3219
WWW: http://www.npes.org

Graphic Arts Technical Foundation
4615 Forbes Ave.
Pittsburgh, PA 15213
Tel.: (412) 621-6941
Fax: (412) 621-3049
E-mail: info@gatf.lm.com
WWW: http://www.gatf.lm.com

Independent Liquid Terminals Association
1133 15th St., NW Suite 650
Washington, DC 20005
Tel.: (202) 659-2301
Fax: (202) 466-4166

Institute of Industrial Engineers (IIE)
25 Technology Park
Norcross, GA 30092
Tel.: (770) 449-0461
Fax: (770) 263-8532
WWW: http://www.iienet.org

International Fabricare Institute (IFI)
12251 Tech Road
Silver Spring, MD 20904
Tel.: (301) 622-1900
Fax: (301) 236-9320
WWW: http://www.ifi.org

International Home Furnishings Marketing Association (IHFMA)
P.O. Box 5687
High Point, NC 27262
Tel.: (910) 889-0203
Fax: (910) 889-7460

Lead Industries Association (LIA)
295 Madison Avenue
New York, NY 10017
Tel.: (212) 578-4750
Fax: (212) 684-7714
E-mail: smith@leadinfo.com
WWW: http://www.leadinfo.com

National Aggregate Association
900 Spring Street
Silver Street, MD 20910
Tel.: (301) 587-1400
Fax: (301) 587-9419

National Association of Printers and Lithographers
780 Palisade Ave
Teaneck, NJ 07666
Tel.: (201) 342-0700
Fax: (201) 692-0286
WWW: http://www.napl.org

National Electronic Manufacturing Association (NEMA)
2101 L Street, NW, Suite 300
Washington, DC 20037
Tel.: (202) 457-8400
Fax: (202) 457-8411
WWW: http://www.nema.org

National Glass Association (NGA)
8200 Greensboro Drive, 3rd floor
McLean, VA 22102
Tel.: (703) 442-4890
Fax: (703) 442-0603
WWW: http://ourworld.compuserve.com/homepages

National Mining Association
1130 17th Street
Washington, DC 20036-4677
Tel.: (202) 861-2800
Fax: (202) 463-6152
E-mail: nma2prime@planetcom.com
WWW: http://www.nma.org

National Paints and Coatings Association
1500 Rhode Island Ave, NW
Washington, DC 20005
Tel.: (202) 462-6272
Fax: (202) 462-8549
E-mail: npca@paint.org
WWW: http://www.paint.org

National Petroleum Refiners Association
1899 L Street, NW 1000
Washington, DC 20036
Tel.: (202) 457-0480
Fax: (202) 457-0486

E-mail: npra3@cais.com
WWW: http://www.npradc.org

National Tank Truck Carriers
2200 Mill Road
Alexandria, VA 22314
Tel.: (703) 838-1960
Fax: (703) 684-5753
WWW:http://www.trucking.org/federation/conferences/nttc.html

Neighborhood Cleaners Association (NCA)
252 West 29th Street
New York, NY 10001-5201
Tel.: (212) 967-3002
Fax: (212) 967-2240
E-mail: ncai@sprynet.com
WWW: http://home.sprynet.com/sprynet/ncai

Paint, Body, and Equipment Association (PBEA)
c/o Martin Fromm and Associates
9140 Ward Parkway, Suite 200
Kansas City, MO 64114
Tel.: (816) 444-3500
Fax: (816) 444-0330
WWW: http://www.pbea.org/

Paper Industry Management Association
2400 E. Oakton Street
Arlington Heights, IL 60005
Tel.: (847) 956-0250
Fax: (847) 956-0520
E-mail: pimastof@wwa.com

Plastic Pipe and Fittings Association (PPFA)
Building C, Suite 20
800 Roosevelt Road

Glen Ellyn, IL 60137
Tel.: (708) 858-6540
Fax: (630) 790-3095

Printing Industries of America
100 Dangerfield Road
Alexandria, VA 22314
Tel.: (703) 519-8100
Fax: (703) 548-3227
WWW: http://www.printing.org

Rubber Manufacturers Association (RMA)
1400 K Street, NW
Washington, DC 20005
Tel.: (202) 682-4800
Fax: (202) 682-4854
WWW: http://www.rma.org

Screen Printing and Graphic Imaging Association International (SGIA)
10015 Main Street
Fairfax, VA 22031
Tel.: (703) 385-1335
Fax: (703) 273-0456
E-mail: sgia@sgia.org
WWW: http://www.sgia.org

Society for Mining, Metallurgy, and Exploration, Inc. (SME, Inc.)
PO Box 625005
Littleton, CO 80162
Tel.: (303) 973-9550
Fax: (303) 973-3845
E-mail: smenet@aol.com
WWW: http://www.smenet.org

Society of Plastic Engineers (SPE)
14 Fairfield Drive
Brookfield, CT 06804-0403
Tel.: (203) 775-0471
Fax: (203) 775-8490
WWW: http://www.4spe.org

Steel Manufacturers' Association (SMA)
1730 Rhode Island Avenue, NW, Suite 907
Washington, DC 20036-3101
Tel.: (202) 296-1515
Fax: (202) 296-2506
E-mail: steelnet@aol.com
cbechak@access.digex.net
WWW: http://www.steelnet.org

Tube and Pipe Fabricators Association, International (TPFA)
5411 East State Street
Rockford, IL 61108
Tel.: (815) 399-8700
Fax: (815) 399-7279

Glossary

Definition of Terms

Batch dyeing Dyeing processes in which a given amount of textile material, usually 100 to 1000 kg, is loaded into a dyeing machine and brought to equilibrium (or near equilibrium) with a solution containing dye.

Biochemical oxygen demand (BOD) A standardized means of estimating the degree of contamination of water supplies, especially those that receive contamination from sewage and industrial wastes. It is expressed as the quantity of dissolved oxygen (in mg/L) required during stabilization of the decomposable organic matter by aerobic biochemical action. Determination of this quantity is accomplished by diluting suitable portions of the sample with water saturated with oxygen and measuring the dissolved oxygen in the mixture both immediately and after a period of incubation, usually five days.

Caustic Any strongly alkaline material that has a corrosive or irritating effect on living tissue.

Chlorofluorocarbon (CFC) Any of several compounds comprised of carbon, fluorine, chlorine, and, sometimes, hydrogen, the best known of which are trichlorofluoromethane and dichlorodifluoromethane. Their use was prohibited in 1979, except for a few specialized items, because of their depleting effect on stratospheric ozone.

Continuous dyeing Processes in which textile materials are fed continuously into a dye "range" at speeds of roughly 50 to 250 meters per minute. Continuous dyeing processes usually consist of dye application, dye fixation with chemicals, or heat and washing.

Dioxin The commonly accepted, though chemically imprecise, name for the compound 2,3,7,8-tetrachlorodibenzo-p-dioxin (TCDD), which is only one of more than 70 members of the family of chlorinated dioxins. It was found to be a contaminant of the herbicide 2,3,5-T (trichlorophenoxyacetic acid). Synthesized in 1957, dioxin is a white, crystalline solid in pure form. It is a carcinogen, a teratogen, and a mutagen. Where soil contamination occurs, the concentration is usually no greater than several parts per million at most. Wastes contaminated with dioxin must be disposed of in officially approved landfills.

Effluent The aqueous mixture from secondary biological waste treatment facilities that enters a stream, river, or other body of water, or such material from a textile operation or pretreatment system entering a sewer for treatment at a publicly owned treatment works (POTW).

Exhaust Dye migration from a dyebath into a fiber or other textile material.

Non-Product Output (NPO) All hazardous substances or hazardous wastes that are generated prior to storage, recycling, treatment, control, or disposal and that are not intended for use as a product.

Pollution Prevention (P2) Changes in production technologies, raw materials, or products that result in the reduction of the demand for hazardous substances per unit of product manufactured and the creation of hazardous products or nonproduct outputs; or changes in the use of raw materials, products, or production technologies that result in the reduction of the input use of hazardous substances and the creation of hazardous byproducts or destructive results; or on-site facility changes in production processes, products, or the use of substitute raw materials that result in the reduction of the amount of hazardous waste generated and disposed of on the land or hazardous substances discharged into the air or water per unit of product manufactured prior to treatment, and that reduce or eliminate, without shifting, the risks that the use of hazardous substances at an industrial facility pose to employees, consumers, and the environment and human health. "Pollution prevention shall include, but need not be limited to, raw material substitution, product reformulation, production process redesign or modification, in-process recycling, and improved operation and maintenance of production process equipment. Pollution prevention shall not include any action or change entailing a substitution of one hazardous substance, product or nonproduct output for another that results in the creation of substantial new risk, and shall not include treatment, increased pollution control, out-of-process recycling, or incineration, except as otherwise provided pursuant to subsection f. of section 7 of this act" (New Jersey Pollution Prevention Act, N.J.S.A. 13:1D-45).

Release Emissions to one or more media, including air, surface water, and land (on-site.) Also included are quantities transferred to publicly owned treatment works (POTWs), and quantities transferred to other off-site locations.

Releases to Air (Point and Fugitive Air Emissions) Includes all air emissions from industry activity. Point emissions occur through confined air streams as found in stacks, ducts, or pipes. Fugitive emissions include losses from equipment leaks, or evaporative losses from impoundments, spills, or leaks.

Releases to Land Include disposal of waste to on-site landfills, waste that is land treated or incorporated into soil, surface impoundments, spills, leaks, or waste piles. These activities must occur within the facility's boundaries for inclusion in this category.

Releases to Water (Surface Water Discharges) Encompass any releases going directly to streams, rivers, lakes, oceans, or other bodies of water. Any estimates for stormwater runoff and nonpoint losses must also be included.

SIC Code The Standard Industrial Classification (SIC) is a statistical classification standard used for all establishment-based federal economic statistics. The SIC codes facilitate comparisons between facility and industry data.

Spent material Any material that has been used and as a result of contamination can no longer serve the purpose for which it was produced without processing.

Strip To remove color from a dyed fabric. This is usually a repair procedure and generally degrades the cloth, especially if cotton.

Substrate Any material upon which ink is impressed, such as paper, plastic, wood, or metal.

Transfers A transfer of toxic chemicals in wastes to a facility that is geographically or physically separate from the facility reporting under TRI. The quantities reported represent a movement of the chemical away from the reporting facility. Except for off-site transfers for disposal, these quantities do not necessarily represent entry of the chemical into the environment.

Transfers to disposal Wastes taken to another facility for disposal generally as a release to land or as an injection underground.

Transfers to energy recovery Wastes combusted off-site in industrial furnaces for energy recovery. Treatment of a chemical by incineration is not considered to be energy recovery.

Transfers to POTWs Wastewaters transferred through pipes or sewers to a publicly owned treatment works (POTW). Treatment and chemical removal depend on the chemical's nature and treatment methods used. Chemicals not treated or destroyed by the POTW are generally released to surface waters or landfilled within the sludge.

Transfers to recycling Sent off-site for the purposes of regenerating or recovering still valuable materials. Once these chemicals have been recycled, they may be returned to the originating facility or sold commercially.

Transfers to treatment Wastes moved off-site for either neutralization, incineration, biological destruction, or physical separation. In some cases the chemicals are not destroyed but are prepared for further waste management.

TRI facilities Manufacturing facilities that have ten or more full-time employees and are above established chemical throughput thresholds. Manufacturing facilities are defined as facilities in Standard Industrial Classification primary codes 20–39. Facilities must submit estimates for all chemicals that are on the EPA's defined list and are above throughput thresholds.

Underground injection A contained release of a fluid into a subsurface well for the purpose of waste disposal.

Use The amount of a hazardous substance actually passing through the facility, or throughput, per year. It is calculated by adding starting inventory of a substance, quantity brought on-site, and quantity produced on-site, and subtracting ending inventory. Use is also equal to the sum of the following quantities: consumed, shipped as or in product, and total Non-Product Output (NPO).

Volatility The ease with which a liquid or solid passes into its vapor state.

Waste ink The ink removed from the ink fountain at the end of a run or following its contamination.

Waste minimization A reduction in the total quantity and/or toxicity of the hazardous waste generated. A waste can be reduced or eliminated at its source, or used, reused, or reclaimed once it is generated.

NOMENCLATURE

AFS	AIRS Facility Subsystem (CAA database)
AIRS	Aerometric Information Retrieval System (CAA database)
AOX	Adsorbable Organic Halides
API	American Petroleum Institute
AQ	Anthraquinone
BDAT	Best Demonstrated Available Technology
BEHP	bis (2-ethylhexyl) phthalate
BIFs	Boilers and Industrial Furnaces (RCRA)
BLM	Bureau of Land Management
BOD_5	Biological Oxygen Demand (5-day), (mg/l)
BOF	Basic Oxygen Furnace
BOP	Basic Oxygen Process
CAA	Clean Air Act
CAAA	Clean Air Act Amendments of 1990

CERCLA	Comprehensive Environmental Response, Compensation and Liability Act
CERCLIS	CERCLA Information System
CFC	Chlorofluorocarbon
CIP	Carbon-in-Pulp
CMA	Calcium Magnesium Acetate
CO	Carbon Monoxide
COD	Chemical Oxygen Demand
CSI	Common Sense Initiative
CWA	Clean Water Act
DAF	Dissolved Air Flotation
D&B	Dun and Bradstreet Marketing Index
DBD	Dibenzo-p-dioxin
DBF	Dibenzofuran
DC	Direct Current
DFE	Design for Environment
DHS	Department of Health Services
EAF	Electric Arc Furnace
E&B	Extraction & Beneficiation (waste)
EB	Electron Beam
ELP	Environmental Leadership Program
EMCC	Extended Modified Continuous Cooking
EPCRA	Emergency Planning and Community Right-to-Know Act
FIFRA	Federal Insecticide, Fungicide, and Rodenticide Act

FINDS	Facility Indexing System
H_2S	Hydrogen Sulfide
HAPs	Hazardous Air Pollutants (CAA)
HDPE	High-Density Polyethylene
HSDB	Hazardous Substances Data Bank
HTMR	High-Temperature Metal Recovery
HVLP	High-Volume-Low-Pressure
IDEA	Integrated Data for Enforcement Analysis
IPA	Isopropyl Alcohol
LDAR	Leak Detection And Repair
LDPE	Low-Density Polyethylene
LDR	Land Disposal Restrictions (RCRA)
LEPCs	Local Emergency Planning Committees
MACT	Maximum Achievable Control Technology (CAA)
MBI	Methylene Bis(phenylisocyanate)
MCC	Modified Continuous Cooking
MCLs	Maximum Contaminant Levels
MCLGs	Maximum Contaminant Level Goals
MDI	Methylene di-p-phenylene isocyanate
MEK	Methyl Ethyl Ketone
METH	Methylene Chloride
MIBK	Methyl Isobutyl Ketone
MSDS	Material Safety Data Sheets
NAAQS	National Ambient Air Quality Standards (CAA)
NaCL	Sodium Chloride

NAE	National Academy of Engineers
NAFTA	North American Free Trade Agreement
NCDB	National Compliance Database (for TSCA, FIFRA, EPCRA)
NCG	Noncondensible Gas
NCP	National Oil and Hazardous Substances Pollution Contingency Plan
nd	No Date
NEIC	National Enforcement Investigation Center
NESHAP	National Emission Standards for Hazardous Air Pollutants
NH_3	Ammonia
NJDEPE	New Jersey Department of Environmental Protection and Energy
NO	Nitrogen Oxide
NO_2	Nitrogen Dioxide
NOV	Notice of Violation
NPDES	National Pollution Discharge Elimination System (CWA)
NPL	National Priorities List
NPO	Non-Product Output
NRC	National Response Center
NSPS	New Source Performance Standards (CAA)
OAR	Office of Air and Radiation
OD	Oxygen Delignification
ODC	Ozone Depleting Chemical
OECA	Office of Enforcement and Compliance Assurance

OPA	Oil Pollution Act
OPPTS	Office of Prevention, Pesticides, and Toxic Substances
OSHA	Occupational Safety and Health Administration
OSW	Office of Solid Waste
OSWER	Office of Solid Waste and Emergency Response
OW	Office of Water
P2	Pollution Prevention
PAH	Polyaromatic Hydrocarbons
PAPRICAN	Pulp and Paper Research Institute of Canada
PCS	Permit Compliance System (CWA Database)
PERC	Perchloroethylene
PIA	Printing Industries of America
PM	Particulate Matter
POTW	Publicly Owned Treatment Works
PPA	Pollution Prevention Act
PWB	Printed Wiring Boards
Q-BOP	Quelle Basic Oxygen Process
RCRA	Resource Conservation and Recovery Act
RCRIS	RCRA Information System
RDH	Rapid Displacement Heating
RMP	Refiner Mechanical Pulping
ROI	Return on Investment
SAGE	Solvent Alternative Guide
SARA	Superfund Amendments and Reauthorization Act

SDWA	Safe Drinking Water Act
SEPs	Supplementary Environmental Projects
SERCs	State Emergency Response Commissions
SEWE	Southeast Waste Exchange
SIC	Standard Industrial Classification Code
SLC	Suppressed Lignin Condensation
SO_2	Sulfur Dioxide
SOP	Standard Operating Procedures
SO_x	Sulfur Oxides
SRP	Salt Recovery Process
STFI	Swedish Pulp and Paper Research Institute
SX-EW	Solvent Extraction and Electrowinning
TCA	Trichloroethane
TCDD	Tetrachlorodibenzo-p-dioxin
TCE	Trichloroethylene
TCF	Totally Chlorine-Free technology
TCRIS	Toxic Chemical Release Inventory System
TMP	Thermomechanical Pulping
TOC	Total Organic Carbon
TRI	Toxic Release Inventory
TRIS	Toxic Release Inventory System
TRS	Total Reduced Sulfur
TSCA	Toxic Substances Control Act
TSS	Total Suspended Solids
TURI	Toxics Use Reduction Institute

UF	Ultrafiltration
USEPA	United States Environmental Protection Agency
UST	Underground Storage Tanks (RCRA)
USUIC	Underground Injection Control (SDWA)
UV	Ultraviolet (radiation)
VOC	Volatile Organic Compound
WGA	Western Governor's Association
WCED	World Commission on Environment and Development
WPC	Whey Protein Concentrate
WWW	World Wide Web

Select Bibliography

"Alaska Health Project." Alaska Department of Environmental Conservation, 1988–89.

Allegri, T.H., Sr. "Handling and Management of Hazardous Materials and Wastes." New York: Chapman and Hall, 1986.

"Amoco—USEPA Pollution Prevention Project." Yorktown, VA: USEPA, Project Summary, January 1992.

"Assessment of Atmospheric Emissions from Petroleum Refining." Austin, TX, and Washington, DC: USEPA, Office of Research and Development, April 1980.

Avidan, A.A., Edwards, M., and Owen, H. "Innovative Improvements Highlight FCCs Past and Future." *Oil & Gas Journal*, January 8, 1990.

Balik, J.A., and Koraido, S.M. "Case Study: Identifying Pollution Prevention Options for a Petroleum Refinery," *Pollution Prevention Review*, 1991.

Bergenthal, J. "Dyebath Reuse Tests Verify Quick Payback." *Textile World*, September 1983.

Brown, L. "Chemical Substitution for 1,1,1-Trichloroethane and Methanol in Manufacturing Operations." *16th Annual Hazardous Waste Research Symposium, 4/90*; EPA/600/9-90/037, August 1990.

Bunshah, R.F. "*Deposition Technologies for Films and Coatings Developments and Applications.*" Park Ridge, NJ: Noyes Publications, 1982.

Bureau of the Census. "1987 Census of Manufacturers Industry Series 34A: Metal Cans, Cutlery, Handtools." U.S. Department of Commerce, April 1990.

———. "1987 Census of Manufacturers Industry Series 34B: Heating Apparatus and Plumbing Fixtures." U.S. Department of Commerce, April 1990.

———. "1987 Census of Manufacturers Industry Series 34C: Fabricated Structural Metal Products." U.S. Department of Commerce, April 1990.

———. "1987 Census of Manufacturers Industry Series 34D: Screw Machine Products." U.S. Department of Commerce, April 1990.

———. "1992 Census of Manufacturers Preliminary Report Industry Series: Petroleum and Coal Products." U.S. Department of Commerce, June 1994.

Callahan, M. "Waste Minimization in Metal Parts Cleaning" *OSWER EPA*; EPA/530-SW-89-049, August 1989.

Calvert, S., and Englund, H. *Handbook of Air Pollution Technology.* New York: John Wiley and Sons, 1984.

Carawan, R. "Pollution Prevention Pays in Food Processing." *North Carolina Agricultural Extension Service*, 1988–89.

Carr, W., and Cook, F. "Savings in Dyebath Reuse Depend on Variations in Impurity Concentrations." *Textile Chemists and Colorists*, May 1980.

"Case Studies of Minimization of Cyanide Waste from Electroplating Operations." USEPA 68-01-7053, December 1986.

"Clean Air Act Complicates Refinery Planning." *Oil & Gas Journal*, May 1991.

"Clean Air Amendments Put Big Burden on Refinery Planners." *Oil & Gas Journal*, June 10, 1991.

"Compilation of Air Pollutant Emission Factors Stationary Point and Area Sources." EPA AP-42, September 1985.

"Construction Materials." *U.S. Industrial Outlook*, 1994.

Cook, F. "Direct Dyebath Reuse: The Future is Now!" *Textile World*, September 1983.

———. "Plant Trials on Dyebath Reuse Show Savings in Energy, Water, Dyes and Chemicals." *Textile Chemists and Colorists*, January 1980.

Cook, F., and Lincher, W. "Dyebath Reuse in Batch Dyeing." *Textile Chemists and Colorists*, January 1978.

Corbgett, R.E. "New Catalyst Designs Meet Environmental Challenges of the 1990's." *Oil & Gas Journal*, October 1, 1990.

Cosgrove, W.J. "Water Pollution and the Textile Industry," *Canadian Textile Journal*, January 1970.

"Costs to the Petroleum Industry of Major New and Future Federal Government Environmental Requirements." Washington, DC: American Petroleum Institute, October 1993.

"Development Document for Effluent Limitations: Guidelines and Standards for the Metal Finishing Industry." *EPA Office of Water Regulations and Standards*, EPA-440-1-80-091A, 1980.

"Development Document for Effluent Limitations Guidelines and Standards for the Textile Mills." USEPA440/1-79/0226, October, 1979.

"Dioxin and Furans—A Primer: What They Are and How to Measure Them." Washington, DC: American Petroleum Institute, March 1990.

Douglas, James M. *Conceptual Design of Chemical Processes*. New York: McGraw-Hill, 1988.

Durney, L.J. *Electroplating Engineering Handbook*, 4th ed. New York: Van Nostrand Reinhold, 1984.

Eckenfelder, W.W., Jr. "Water Quality Engineering for Practicing Engineers." New York: Barnes & Noble, 1970.

Edde, H. "Environmental Control for Pulp and Paper Mills." Park Ridge, NJ: Noyes Publications, 1984.

"Emissions from Metal Finishing Operations." Draft report, USEPA Office of Research and Development, March 31, 1995.

Englehardt, J.D. *Journal of Hazardous Material*, Amsterdam: Elsevier Science Publishers B.V., 1993.

"Environmental Design Considerations for Petroleum Refining Crude Processing Units." American Petroleum Institute, February 1993.

"EPA Facility Pollution Prevention Guide." EPA/600/R-92/088, Office of Solid Waste USEPA, 1992.

"EPA Profile of the Electronics and Computer Industry." USEPA EPA 310-R-95-002, September 1995.

"EPA Profile of the Inorganic Chemical Industry." USEPA EPA 310-R-95-004, September 1995.

"EPA Profile of the Iron and Steel Industry." USEPA EPA 310-R-95-005, September 1995.

"EPA Profile of the Metal Mining Industry." USEPA EPA 310-R-95-008, September 1995.

"Estimating Chemical Releases from Electroplating Operations." USEPA OPPTS Title III, Section 313, Release Reporting Guidance, 1988.

Galecki, G. "Proceedings of Waste Stream Minimization and Utilization Innovative Concepts An Experimental Technology Exchange." *Battelle Memorial Institute*, PNL-SA-19337, April 1991.

Gardener, Deirde and Borne, B.J. *"Textile Wastewaters: Treatment and Environmental Effects." Journal of the Society of Dyers and Colourists*, August 1978.

"The Generation and Management of Wastes and Secondary Materials in the Petroleum Refining Industry." *American Petroleum Institute*, February 1991.

"Generation and Management of Wastes and Secondary Materials: Petroleum Refining Performance." *American Petroleum Institute*, Survey, June 1992.

Giese, J. "Sanitation: The Key to Food Safety and Public Health." *Food Technology*, December 1991.

"Guidance Manual for Electroplating and Metal Finishing Pretreatment Standards." USEPA/Effluent Guidelines Division and Permits Division, 1984.

"Guides to Pollution Prevention: The Commercial Printing Industry." Risk Reduction Engineering Laboratory and Center for Environmental Research Information, EPA/625/7-90/008, August 1990.

"Guides to Pollution Prevention: The Metal Finishing Industry." USEPA, ORD, October 1992.

Hardy, J. "Dairy Cuts Wastewater One-Third." *Food Processing*, August 1991.

Hawkins, R.F. "Use of Water-Based Inks to Meet EPA Requirements." Paper 87-48.4. 80th Annual Air Pollution Control Association Meeting and Exhibition, New York, June 21–26, 1987.

"Hazardous Waste Generation: 1. Petroleum Refining." USEPA, Office of Solid Waste, January 1994.

Hazelwood, D. "Program Toward Waste Minimization in Painting Operations in the Defense Community." International Conference on Pollution Prevention; EPA/600/9-90/039, September, 1990.

Huisingh, D., Hilger, H., Thesen, S., and Martin, L. "Profits of Pollution Prevention" *A Compendium of North Carolina*

Case Studies. North Carolina Department of Natural Resources and Community Development, May 1985.

Hunt, R.G. "Dyebath and Bleach Bath Reconstitution for Textile Mills." *Pollution Prevention Pays Tips*. Raleigh, NC: North Carolina Department of Natural Resources, 1985.

Hunt, R.G., and Chestnut, S. "Industrial Resource Recovery Practices: Textile Product Industries." USEPA, June 1982.

"Improved Dyeing System Reduces Energy, Chemical and Labor Requirements for 100% Cotton and Natural Blend Knit Fabrics." *Cotton Incorporated News Bulletin*, May 1, 1977.

"Industrial Surveys," vol. 2. New York: Standard and Poor's, 1995.

"Industry Profile for the Metal Finishing Industry." Meridian Research, USEPA/OPPT, June 24, 1994.

"Issues Affecting the Refining Sector of the Petroleum Industry." *Hearings Before the Committee on Energy and Natural Resources, United States Senate*, Washington, DC, May 19, 1992.

Jacobs Engineering Group for Alternative Technology Section. "DHS Waste Audit Study: Fabricated Metal Products Industry." Toxic Substances Control Division, California Dept. of Health Services, 1989.

Jacobs Engineering Group. "Guides to Pollution Prevention: The Fabricated Metal Products Industry." ORD; EPA; EPA/625/7-90/006, July 1990.

Jones, H. "Pollution Control in Meat, Poultry and Seafood Processing." Noyes Data, 1974.

———. "Waste Disposal Control in the Fruit and Vegetable Industry." Noyes Data, 1973.

Klink, K., Cushnie, G., Gallerani, P., and Higgins, T. "Metal Plating and Surface Finishing." In *Pollution Prevention Handbook*, ed. Thomas E. Higgins. London: CRC Lewis Publishers, 1995.

Kohl, J., Moses, P., and Triplett, B. *Management Recycling Solvents: North Carolina Practices, Facilities, and Regulations.* Raleigh, NC: North Carolina State University, 1984.

Konstandt, F. *Organic Coatings Properties and Evaluation.* New York: Chemical Publishing Co., 1985.

Kulube, H. "Residual Carrier Components in Exhausted Textile Dyebaths." Master's thesis, North Carolina State University, 1987.

Leeman, J.E. "Hazardous Waste Minimization: Part V Waste Minimization in the Petroleum Industry." *Journal Air Pollution Control Association* 38, June 1988.

Leeper, S. "Production of Organic Chemicals via Bioconversion." *EG&G Idaho,* for Idaho National Engineering Laboratory, DOE, July 1991.

Leffler, W.L. *Petroleum Refining for the Non-Technical Person.* Tulsa, OK: PennWell Publishing Company, 1985.

Licis, L. "Industrial Pollution Prevention Opportunities for the 1990s." EPA/600/8/-91/052, August 1991.

Lund, H. *Industrial Pollution Control Hand Book.* New York: McGraw Hill, 1971.

"Machining, American Society for Metals." *Metals Handbook* vol. 9, no. 16, 1989.

Meyers, R.A. *Handbook of Petroleum Refining Processes.* New York: McGraw-Hill, 1986.

Nadeau, T., and Dejak, M. "Copper, Nickel and Chromium Recovery in a Jobshop." *Plating and Surface Finishing,* April 1986.

"NITS Document PB84206465." Washington, DC: U.S. Department of Commerce, May 1984.

Norman, M.E., Kapoor, S., Smalley, G.A., and Daniel, B.M. "U.S. Regs Cause Refiners to Rethink Wastewater Systems." *Oil & Gas Journal,* June 1, 1992.

Noyes, R. *Pollution Prevention Technology Handbook*. Park Ridge, NJ: Noyes Publications, 1993.

"Open Width Pad/Batch Dyeing of Cotton Fabrics." *Cotton Incorporated Technical Bulletin TCR 207*, December 1978.

"Petroleum: An Energy Profile." Energy Information Administration, August 1991.

"Petroleum Refinery Enforcement Manual." Arlington, TX: USEPA Office of Enforcement by PEDCO Environmental, March 1980.

Petty, J. "Low Energy Dyeing of Type 6 Nylon Carpet Yarn." *American Dyestuff Reporter*, June 1981.

Pferdehirt, W. "Roll the Presses But Hold the Wastes: P2 and the Printing Industry." *Pollution Prevention Review*, Autumn 1993.

Pinkerton, H.L., and Graham, K.A. "Rinsing." *Electroplating Engineering Handbook*, 4th ed. New York: Van Nostrand Reinhold, 1984.

"Pollution Prevention Bibliography." *North Carolina Department of Natural Resources and Community Development*, January 1986.

"Pollution Prevention in Metal Manufacturing: Saving Money Through Pollution Prevention." USEPA, OSW, October 1989.

"Pollution Prevention Opportunities Checklists." *County Sanitation Districts of Los Angeles County*, January 1991.

"Pollution Prevention Opportunities in Paper Manufacturing." EPA, October 1990.

"Pollution Prevention Opportunities in Petroleum Refining (Fact Sheet)." Philadelphia, PA: USEPA Region III, October 1990.

"Pollution Prevention Options in Metal Fabricated Products Industries: A Bibliographic Report." USEPA OPPT, January 1992.

"Profile of the Electronics and Computer Industry." USEPA, Office of Compliance Sector Notebook, EPA 310-R-95-002, 1995.

"Profile of the Pulp and Paper Industry." USEPA, Office of Compliance Sector Notebook, EPA 310R-95-015, 1995.

"Profile of the Wood Furniture and Fixture Industry." USEPA, Office of Compliance Sector Notebook, EPA 310-R-95-003, 1995.

"Prolonging Machine Coolant Life." *Minnesota Technical Assistance Program*, nd.

"Properties and Selection: Stainless Steels, Tool Materials and Special Purpose Materials." *Metals Handbook*, American Society for Metals, 1980.

Queen, L. "Size Recovery Using Ultrafiltration Membranes." *Textile Industries*, August 1981.

Ragsdale, R. "U.S. Refiners Choosing Variety of Routes to Produce Clean Fuels." *Oil & Gas Journal*, March 21, 1994.

Ramsey, M. "Processing Rice Bran by Supercritical Fluid Extraction." *Food Technology*, November 1991.

"Refinery Wastewater Priority Pollutant Study—Sample Analysis and Evaluation of Data." Washington, DC: American Petroleum Institute, December 1981.

Roderick, G.W., and Nastri, D. "Dry Scrubber Reduces SO_2 in Calciner Flue Gas, Brown." *Oil & Gas Journal*, February 18, 1991.

Rubey, Robert G. "Wastewater Heat Recovery." Midwinter Conference on Textile Wastewater Treatment and Air Pollution Control, Hilton Head Island, SC, January 1982.

"Russell's Condensate System Saves $1,000/day." *Textile World*, June 1986.

Schaffer, G. "Recycling Coolant Reduces Cost." *American Machinist*, February 1978.

Schwartz, S.I. "Recycling of Hazardous Waste Solvents: Economic and Policy Aspects." *Solvent Waste Reduction Alternatives Symposia Conference Proceedings*, Los Angeles, 1986.

"Selection of Cleaning Process Metals." *American Society for Metals Committee on Selection of Cleaning Process Handbook* 9, nd.

Shen, T.T. *Industrial Pollution Prevention*. New York: Springer-Verlag, 1995.

Sluhan, W.A. "The Application of High Speed, Disc Bowl Centrifuges to Water Miscible Cutting and Grinding Fluids." Perrysburg, OH: Master Chemical Corporation, nd.

Smith, Brent, "Determining Optimum Wet Pickup Levels in Low Add-On Finishing." *American Dyestuff Reporter*, May 1985.

———. "Identification and Reduction of Pollution Sources in Textile Wet Processing." *Pollution Prevention Pays Program*. NC Division of Environmental Management, 1986.

———. "Solid Waste Disposal—A Case History." *Textile Wastewater Treatment and Air Pollution Control Conference*, Hilton Head Island, SC, February 1981.

———. "Workbook for Pollution Prevention by Source Reduction in Textile Wet Processing." *Pollution Prevention Pays Program*. NC Division of Environmental Management, 1988.

Smith, B., and Rucker, J. "Water and Textile Wet Processing." *American Dyestuff Reporter*, July 1987.

Smook, Gary A. "Handbook for Pulp & Paper Technologists." Vancouver, BC: Angus Wilde Publications, 1992.

Sokolowska-Gajda, J., Freeman, H.S., and Reife, A. "Synthetic Dyes Based on Environmental Considerations." Reprinted from *Textile Research Journal* 64, July 1994.

Solid Waste Management Newsletter. Chicago: University of Illinois, vol. 6, no. 3, March 1992.

"Solvent Substitution." First Annual International Workshop on Solvent Substitution, 1992.

Stewart, J.M. "Aquatic Bioassay Toxicological Summary." January 15, 1986.

Stone, R.L. "A Conservative Approach to Dyeing Cotton." *Cotton Incorporated,* May 1979.

"Strategy for the Regulation of Discharges of PHPDs and PHDFs from Pulp and Paper Mills to Waters of the United States." EPA, May 21, 1990.

"Success Through Waste Reduction." Washington State Department of Ecology, Publication 90-22, 1990.

Suchecki, Stanley M. "Troubleshooting in the Dyehouse." *Textile Industries,* December 1977.

"Summary of Technologies for the Control and Reduction of Chlorinated Organics from the Bleached Chemical Pulping Subcategories of the Pulp and Paper Industry." EPA, April 27, 1990.

"Surface Cleaning, Finishing, and Coating, American Society for Metals." *Metals Handbook* 5, 1982.

"Sustainable Environmental Law." *Environmental Law Institute.* St. Paul, MN: West Publishing Co., 1993.

"Sustainable Industry: Promoting Strategic Environmental Protection in the Industrial Sector." *Phase 1 Report,* USEPA, OPPE, June 1994.

Tincher, W. "Energy Conservation in Carpet Dyeing by Dyebath Recycling." *American Dyestuff Reporter,* May 1977.

Tincher, W., et al. "Reusing Dyebaths in Jet Drying." *Textile Chemists and Colorists,* December 1981.

"Toxic Chemical Release Inventory: Clarification and Guidance for the Metal Fabrication Industry." U.S. EPA, OTS, 1990.

"Toxics Release Inventory, Public Data Release." U.S. EPA, Office of Pollution Prevention and Toxics, April 1994.

"The U.S. Petroleum Industry: Past as Prologue." *Energy Information Administration*, September 1993.

"U.S. Petroleum Refining: Meeting Requirements for Cleaner Fuels and Refineries," I–VI. National Petroleum Council Committee on Refining, U.S. Department of Energy, August 1993.

"Waste Minimization in the Petroleum Industry: A Compendium of Practices." *American Petroleum Institute*, November 1991.

"Waste Minimization Opportunity Assessment Manual, EPA." Cincinnati: Hazardous Waste Engineering Research Laboratory, EPA/625/7-88/003, April 1988.

Watkins, R.N. *Petroleum Refining Distillation*. Houston, TX, Gulf Publishing, 1979.

Wentz, C.A. *Hazardous Waste Management*. Argone National Laboratory. New York: McGraw-Hill, 1989.

Williams, B. "Environmental Related Issues Taking Their Turn in Restructuring Industry." *Oil & Gas Journal*, January 22, 1990.

Willson, E. "Recycling/Waste Minimization: One Mill's Success Story." *Technical Association of the Pulp and Paper Industry Proceedings*. Environmental Conference Atlanta, GA: TAPPI Press, 1991.

Wolf, I. "Critical Issues in Food Safety." *Food Technology*, January 1992.

Yang, S. "Proceedings of Waste Stream Minimization and Utilization Innovative Concepts: An Experimental Technology Exchange." *Battelle Memorial Institute,* PNL-SA-19337, April 1991.

Yazdanl, A. "Regulatory Pressures on Chlorinated Solvents: How Can Users Cope." *AESF/EPA Proceedings,* January 27, 1992.

Zabik, M.J. "Unique Methods for Extending the Life of Synthetic Metalworking Fluids in a Manufacturing Environment." *ASLE Paper No. 87-AM-6B-3,* Anaheim, CA, May 1987.

The HOW2DOP package is available online at www.wiley.com/go/healey under the "downloads" tab. Simply unzip the archive to a local folder and launch index.htm in Internet Explorer.

This zip file is an archive of an ancillary webpage originally provided on an accompanying disk with "Pollution Prevention Opportunity Assessments: A Practical Guide". The webpage dates from 1999, and due to the manner in which it was written, the navigation frames only work correctly with Microsoft's Internet Explorer. If you experience problems with the ancillary material, you may still view the content by opening the "*.htm" files individually in your browser of choice.

Index

Note: **Boldface** characters indicate Tables and/or Graphs

K

L

M

N

O

R

S

T

U

About the Authors

Marcus J. Healey, Ph.D., M.B.A., is a multi-disciplinary environmental professional and Director of the New Jersey Technical Assistance Program for Industrial Pollution Prevention at New Jersey Institute of Technology. Dr. Healey is an engineer and scientist with twenty-two years of domestic and overseas experience in operations, research, management, and training. Before joining NJIT, Dr. Healey worked in the public and private sectors, both domestically and abroad. His strengths include pollution prevention, biological waste treatment, microbial methods of development, and environmental management information systems.

Daniel Watts, Ph.D., is the Executive Director of the Emission Reduction Research Center and the Deputy Executive Director of the Center for Environmental Engineering and Science, both at NJIT. He also manages the NJIT component of the U.S. Army Sustainable Green Manufacturing Program.

Laura L. Battista has her Master of Science degree in Environmental Policy from NJIT. Ms. Battista is a pollution prevention analyst at NJTAP working with manufacturing facilities in the areas of regulatory compliance, pollution prevention, life cycle analysis, waste management, and policy development.

Douglas Kretkowski graduated from NJIT with a B.S. in Chemical Engineering. Currently, Mr. Kretkowski is a pollution prevention analyst with NJTAP performing site assessments for small and mid-sized manufacturing facilities throughout the state. In addition, Mr. Kretkowski is a member of the NJIT HazMat Response Team and is NJTAP's Webmaster.

Jeffrey Lewis graduated from Princeton University with a B.S.E. in Chemical Engineering. Mr. Lewis was a pollution prevention analyst with NJTAP for two years before moving on to investment banking. Mr. Lewis also has a strong background in various application software.

Jeffrey Pico is a researcher with NJTAP. He performs pollution prevention assessments for small manufacturing companies. Mr. Pico possesses a Bachelor's degree from Cornell University in Operations Research and Industrial Engineering. Prior to joining NJTAP, Mr. Pico served as a management consultant with the New York City Department of Correction.

Thomas Grosvenor was a pollution prevention analyst with NJTAP for over one year. His focus was on regulatory guidance and pollution prevention opportunity assessments. He is currently involved with project and environmental engineering work at Foster Wheeler USA Corporation.

Printed and bound by CPI Group (UK) Ltd, Croydon, CR0 4YY
16/09/2026
14963457-0001